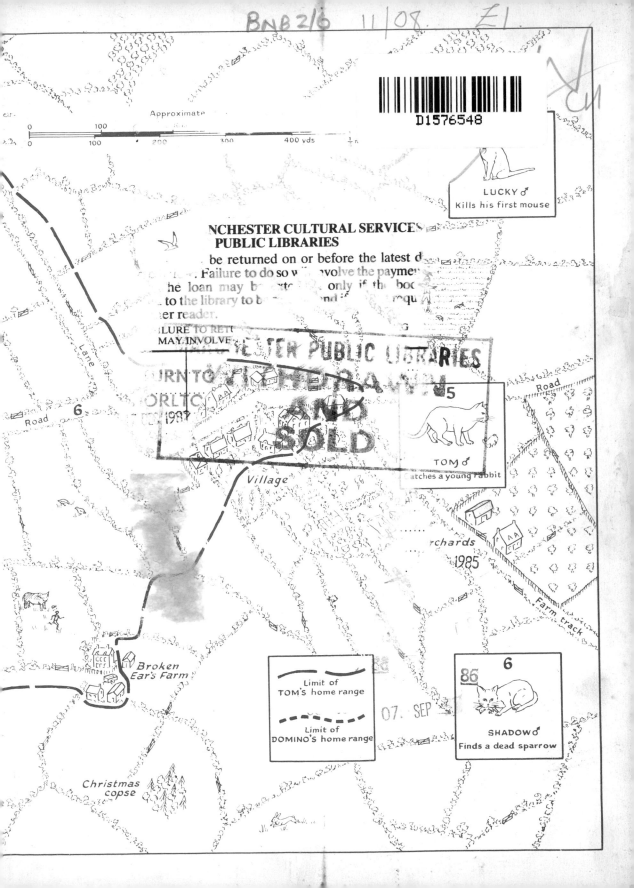

Approximate

LUCKY ♂
Kills his first mouse

5
TOM ♂
atches a young rabbit

Village

rchards

Farm track

6
Road

Lane

Road

6

Broken
Ear's Farm

Christmas
copse

Limit of
TOM'S home range

Limit of
DOMINO'S home range

86
6
SHADOW ♂
Finds a dead sparrow

THE CURIOUS CAT

Michael Allaby
and Peter Crawford

Michael Joseph · London

First published in Great Britain by
Michael Joseph Limited
44 Bedford Square, London W.C.1
1982

ISBN 0 7181 2065 5

Filmset and printed in Great Britain by BAS Printers Limited, Over
Wallop, Hampshire and bound by Hunter and Foulis Ltd., Edinburgh

Contents

Acknowledgements

The authors wish to thank the following for permission to reproduce the photographs used in this book:

Vanellus Productions Ltd., Maurice Tibbles: for those appearing on pages 13, 49, 55, 82, 107 and 118 (*right*)

Radio Times: for those appearing on pages 14, 68, 139 and 144

Peter Apps: for those appearing on pages 20, 36, 41, 42, 46, 97, 110, 118 (*left*), 122, 128, 129, 132, 134 (*both*), 141 and 143

Peter Crawford took the photographs which appear on the following pages: 16, 17, 21, 25, 27, 33, 50, 64, 65, 127, 136, 148, 153, 155

Boris Weltman drew the endpaper map

Introduction

EARLY IN 1978, as one of the harshest winters ever known in that part of the country was promising to turn into a North Devon spring, a small team of people began work in a barn and a hut not far from the Atlantic coast. They were preparing for a year during which they would study and film the behaviour of four cats, who would be allowed to roam free in the countryside. The project ended a year later. The 'end product', as far as most people were concerned, was a film called 'The Curious Cat', which was shown on BBC2 in the series *The World About Us*.

It all began as an idea from Peter Crawford, a producer with the BBC Natural History Unit. For some years Peter, a cat enthusiast, had cherished the notion of filming the behaviour of ordinary cats; for him the private lives of these familiar animals held as much fascination as the exotic wildlife more often featured on the television screen.

The cat has lived with humans for many centuries – though perhaps for a shorter time than some of us imagine – and because it shares our homes and lives in our towns, we see more of it than we see of cattle, say, or sheep. Notoriously, though, the domestic cat insists on retaining much of its independence, and as every cat owner knows it is liable to leave a home that displeases it. What happens to it then? In our cities, such homeless cats live as best they can by catching vermin if they are skilled in hunting, or scavenging as alley cats. In the countryside, though, the status of the cat is rather different. Farm cats are 'employed' to control rodents and they live fairly independent lives about which surprisingly little is known. They are seen when they turn up at the dairy each milking time, and now and then you may catch a glimpse of one as it goes about its business in the fields and along the hedgerows. This study set out to fill in some of the gaps in this picture, by watching a group of such cats closely to see how they fared.

From the very beginning, the project involved collaboration between the film-makers and the scientists who advised them and who used the opportunity to learn more about the way cats live. The central question for everyone concerned the sociability of cats. Is the cat the solitary animal that features in our literature and folklore, or are there circumstances in which several cats will collaborate? In either case, what are the implications for our understanding of the cat family in general and of the position of the cat among carnivorous mammals?

The Devon study was of short duration, the number of cats involved was very small, and predictably the results were tantalisingly inconclusive. Yet aspects of cat behaviour were observed by the scientists that had never been observed in this way before, and they were recorded on film for the first time. Despite its modest scale, the project broke new ground, and served as a focal point for new thinking about the nature of the cat.

The film proved highly popular, but there is a limit to the story any film can tell. In this book, as we tell the story again, we are able to expand, to include the material that could not be included in the film, and to relate what was seen to what we know already about cats and other animals. This is a book about cats, then, but especially it is a book about Tom, Smudge, Pickle, Domino and a very special kitten. In the course of the story we witness most of the events that are important in the life of a cat: mating, the birth of kittens, the 'education' of the kittens as they learn to fend for themselves, hunting, feeding, sleeping, sickness, injury and death. Our farm cats experienced all of these during that one short year. We look, too, at the way a scientific study of animal behaviour is organised and conducted, and at the equipment that is used, and we consider, briefly, the way this film was made.

The film ended with the winter of 1978–79, but the project has continued back at Oxford University, and has flourished into a long-term study of animal behaviour. In this book we tell the story of the family of cats that started it all down in Devon, and we describe their adventures after the filming came to an end. By the summer of 1980 we knew the path that each of the cats would take, most probably for the rest of its life.

The project was financed by the BBC. In the past, the Natural History Unit had sponsored film projects that also yielded useful scientific information, but this was the first time the unit had set up a scientific study specially for a film. Peter Crawford's first task, once his proposal had been accepted, was to find a scientific consultant for the filming – someone who would set up the study project. Peter asked Dr David Macdonald of the Department of Zoology at Oxford University to supervise the scientific aims of the venture, and to appoint a qualified research student who would observe the cats in the wild. David chose Peter Apps, and the cats were selected by the RSPCA. They had been born on a farm in Oxfordshire, and were no longer wanted. The RSPCA had been asked to destroy them, and it was the cats' involvement in this study that ensured their survival in all but one case, so far as we know, to the present day.

Peter Apps and the cats took up residence, in a hut and barn respectively, on the small farm owned by Maurice and Edith Tibbles. Maurice was the cameraman who spent countless hours sitting behind his viewfinder waiting for those moments that made the film so memorable. Without the diligence of Peter and Maurice the film could not have been made, and we would have no story to tell.

Was it all worth while? The Devon Study certainly yielded information that has proved scientifically useful, and there is no doubt that the film also provided popular entertainment and an insight into the wider world of the domestic cat. Beyond that, perhaps you are the best judge. In our opinion it helped in the slow business of piecing together, bit by bit, the details that make up the life of this most familiar, yet most enigmatic, of animals. As cat enthusiasts we may be prejudiced, but we found our glimpses of the private life of the independent cat fascinating and in many ways surprising. We hope you will be fascinated. We suspect you will be surprised.

Michael Allaby   Peter Crawford
Wadebridge, Cornwall   Bristol

Down on the farm

IT WAS early morning, the time that humans associate with breakfast, and it was spring. From the little wooden hut hidden among the rhododendrons there came a smell of frying bacon. The song-birds were shouting defiance at one another from the trees, watched invisibly by the old buzzard who sat immobile, high in an elm, waiting with the patience of a predator for the signal that would mark the start of his hunting day. The air was cold and the plants were sodden with dew, but the clean blue of the sky promised warmth.

Tom made his way through the long, unkempt grass and past the patches of hogweed, keeping as dry as he could. Now and then he would divert to one side or the other, sometimes to pass round an obstacle, sometimes to investigate what might have been an interesting smell, but had you been watching him you would have said there was a purpose in his journeying. He was going somewhere. He did not hurry, for his errand was not urgent, but neither did he dawdle. He looked confident, relaxed, and yet alert. For him the world was new. He was returning from his first exploration of the farm.

Had he been a human, he would have been able to see the bright metal of the barn for which he was heading. As it was, the vegetation hid it from him until he reached the more open ground of the small yard that had been made a few weeks earlier by clearing away undergrowth. The yard had been created partly for Tom's benefit and partly to help Peter Apps, for they were colleagues, the half-wild tomcat and the young zoologist, bound to keep one another company for the year their study project would last.

It was Peter who was frying bacon in his hut. That was where he lived, the hut. To this day, long after the farm project has ended, it is still known as 'Pete's hut'. It began as a fourteen by ten feet prefabricated building, like those you see on building sites. It did not end there, though, for after he had assembled it himself, Peter set about making it habitable. He lagged the roof and lined it and the walls

Tom explores the barn – a new home for the cats

with board. He built shelves for his books and scientific equipment, a bench for working, and cupboards for storage. He installed a sink, though without taps, a gas-cooker and heater, and electric power. He pinned posters to the walls, and above his bed he fixed a picture of a battered Snoopy bearing the legend 'Never trust a smiling cat!'.

From the window of the hut he had a view across the small fields of North Devon and, a little to the right, he could see a heronry.

Even from the open side, the side with the view, the hut was difficult to see. Tall bushes grew around and hid it. Until you were almost upon it, the side with the door was hidden completely by rhododendron bushes. Although close to the barn, Peter's hut was entirely separate. It was important that the cats, who were to be the subjects of the study, did not follow Peter and become pets. So far as they were concerned, he had to be neutral, quite impersonal, an animate but irrelevant feature of the environment that never showed what a cat would understand as interest. For Peter it was not always easy.

He was an ideal choice for the job, this lanky young man with the dark hair, droopy moustache and long, narrow, rather earnest face. He had been brought up on his father's farm in Kent and understood country

Peter Apps sets up home in his hut

life. There had been cats on that farm, of course, and so he knew, in a general kind of way, the place cats occupy on a working farm. No less important, though, his rural background made him acceptable to local people. He was an outsider, for all newcomers are inevitably outsiders in a small, isolated rural community, but he did not intrude or interfere with the important business of the neighbourhood. People became used to him as he passed them on his small motorbike, wearing a bright yellow crash helmet, or as he moved on foot in wellies and a green parka with the hood up, usually dripping wet and often waving a radio aerial like some exotic wand. When his aerial first made its appearance, it caused some local alarm. People knew he had something to do with the BBC, and they also knew a campaign had been launched (in fact by the Post Office, not the BBC) to detect people who were using television sets without a licence. It was some time before they could be reassured. They remember him still.

It was David who sent Peter to Devon. Dr David Macdonald works at the Department of Zoology at Oxford University, where he specialises in the study of animal, and especially carnivore, behaviour. He has observed jackals and other large animals in the Near East, wolves in Italy, monkeys in Borneo, and in this country he is a leading authority on foxes. Some years earlier, he had collaborated with the BBC Natural History Unit in making a film called 'The Night of the Fox'. It was the success of that collaboration which led to his involvement in the study of the farm cats.

This project was to be rather different from most of those that had earned the Natural History Unit its reputation. There had been film projects that had yielded useful scientific information, but essentially they had been exercises in journalism. This time, the aim was to generate information as well as popular entertainment, to combine scientific observation and film-making quite deliberately. So, for this film, the BBC found itself sponsoring and financing scientific research. Initially, the research would be on a modest scale, for its duration would have to be dictated by the filming schedule and the number of animals involved would have to be small, but the science and the filming would complement one another. In one sense, the filming would need the help of the scientists. At least, this is the way it seemed to Peter Crawford, the trained zoologist and television producer whose idea it was. A cat enthusiast himself, Peter had long cherished the notion of making a film about ordinary cats, but as the idea took shape he found there was very little scientific information available to him. This most familiar of animals had been little observed in the business of its everyday life.

The idea appealed to David, for he saw in it the opportunity to study cat behaviour and to relate that to what he knew already about the behaviour of carnivores in general. He wanted to look at the domestic cat as a predator, to try to fit it into the broad mosaic of carnivore behaviour, to discover what kind of animal is beneath the skin of the fireside moggy. To do this, he would adapt the techniques and the technology that had been developed for the study of large game animals, mainly in Africa, and for foxes and badgers in Britain.

He needed a field assistant who would spend a year living with the cats, collecting data that would be processed at Oxford. At that time, Peter Apps had just graduated in zoology. He wanted to enter research, and he had let it be known around the Zoology Department that he was keen to join a research team. He was finding it difficult, because research funds were not easy to obtain, research teams were necessarily small, and competition for places in them was keen. Peter was ideal for the cat project and he jumped at the opportunity.

Meanwhile, the search for the cats intensified. The RSPCA was helping by reporting colonies of cats as its officers heard of them, and David and Peter Crawford spent some time on 'wild cat chases'. Their requirements were fairly precise. The ideal group of cats would be a small natural colony that was already fairly independent of humans, but not completely wild.

The domestic cat, *Felis catus*, has lived with humans for a long time. The first record we have of what appears to be a domesticated cat is in a fragment of the Egyptian Book of the Dead dating from about 1600 BC. By that time most of our other domesticated animals were firmly under human control. The cat was a late arrival in the human household. But the cats that today grace our hearthrugs and back alleys are the descendants of those Egyptian domestic cats. Despite its rather similar appearance, the domestic cat is not the same species as *Felis sylvestris*, the truly wild cat of northern Europe. It can be confusing, then, to talk of 'domestic' cats as 'wild'. We must reserve 'wild' for *F. sylvestris* and the other twenty-five or so *Felis* species around the world. When a domesticated animal takes to living without humans, the usual term for it is 'feral'. For most purposes this is adequate, for it distinguishes between its domesticated ancestors and any truly wild relatives it may have. In the case of the cat, however, it is not enough, for the relationship between cats and humans exists at every stage, from close contact to total avoidance. If the cat that sleeps on your lap is domesticated, in the sense that it is highly, or perhaps totally, dependent on humans, at the other end of the scale there screams, cowers and spits the cat that has forgotten – if it ever knew – that humans might make its life easier. It cannot be handled, even if it can be caught, and it is likely to be heard more often than it is seen, for it will avoid humans and hide at their approach. It may take food that is left for it, but it will not be seen to take it. Under no circumstances will it enter a human dwelling voluntarily. Between the two extremes there is an almost infinite gradation, and it may be defined in terms of the degree of dependence the animal has on human help. 'Feral' is an adjective that is only a little more useful to us than 'wild'.

What the project required was cats whose position on this scale of dependence was known. If the cats were too dependent, their behaviour would not be typical of cats that must fend for themselves. If they were too independent, they might live well enough, but they would seldom be seen, and that would make them very difficult to study and virtually impossible to film. The cats had to be capable of and accustomed to looking after themselves, but it must be possible to handle them occasionally should this prove necessary, and they must tolerate the sight, sound and smell of the human observers.

The RSPCA was reporting to David Macdonald colonies of cats they had been asked to destroy. Such colonies of more or less independent cats are fairly common, and there is not much the RSPCA can do for them. No one will give a home to a full-grown animal that distrusts humans and has never lived in a human home, not when there are plenty of more amenable kittens to be had for the asking. Such cats are destroyed, almost invariably, and were it not for the Devon project, there is not the slightest doubt that our cats would have met the same fate. The disturbance they experienced by being moved to their new home seemed totally justified – it ensured their survival and provided for their welfare.

Cats are protected under British law, which does not recognise the word 'feral' and so does not have to define it. The cat is a domestic animal, and it is an offence to maltreat it. Maltreatment includes the withholding of veterinary treatment where this leads to suffering or death that could have been avoided, and the deliberate withholding of food from an animal that is unable to feed itself. In practical terms, stray cats forfeit this protection, because it is impossible to identify a responsible human who can be made to care for them.

When at last the RSPCA found a group of cats that seemed suitable, David had the job of catching them. The group consisted of one full-grown male, one mature female, and two younger females who appeared by their coat markings to be the daughters of the two older cats. They were – or seemed to be – a family group. One night David caught them in traps and took them to the new home that had been prepared for them down in Devon.

Tom had reached the barn. He was big, tough, and sometimes haughty, a typically battered tomcat who put up with no nonsense from anyone. He did just as he liked, and the world could make the best of it. He would have passed unnoticed in any street. His white fur with its black markings was just like that of millions of pet cats. As he entered the barn, Pickle approached him. She, too, was black and white, but there was some brown in her fur. The two cats met, tails held erect in greeting, touched noses, then sidled past one another, their flanks touching and their tails bent slightly over the back of the other cat, until each reached the tail end of the other, where the formal greeting ended with a cursory sniff. The greeting-cum-inspection completed, Tom began to groom the back of Pickle's neck, gently at

Tom – 'big, tough and haughty'

first and almost absent-mindedly, but then with more passion. The grooming led to a bite on the neck, which held Pickle still as he mounted her. She tucked her legs beneath her, Tom's tail jerked rhythmically, and successful copulation was confirmed as Pickle screamed in pain, a part shriek, part spit, that caused her to twist around and swipe at Tom. The pain appeared to subside at once and the two animals moved apart. Tom had come home, announcing his arrival in the way that was customary when he came across one of the females in oestrus – on heat. It did not trouble either of them that, almost certainly, Pickle was his daughter.

Her pain was real. The male cat has barbs on its penis that permit smooth entry but that stab on withdrawal. It is the stab that causes the shriek, but it also stimulates hormonal activity that causes an egg to be released for fertilisation. The mechanism is subtle, selected by evolution as appropriate for an animal that spends much of its time alone. Is the cat a social animal or a solitary one? Its method of copulation and ovulation might suggest that it is solitary by nature. If the female were to release eggs according to a regular cycle, fertilisation would depend on the coincidence of meeting a male at the precise moment when conception was possible. Where the environment provides food for only a few cats, the hunters must spend most of their time dispersed over a wide area. Individuals meet rarely, and by chance. The copulatory mechanism triggers ovulation and ensures that individuals have a greater chance of producing offspring. Whether the cat is essentially social or solitary by nature was one of the questions that David Macdonald and Peter Apps hoped to understand better by the end of their year spent observing Tom and his female harem.

It was Peter who named the cats. Except for Tom, the names were based on fur markings. Pickle, with her white, black and brown markings, suggested pickle – at least to Peter. Domino was clearly marked in black and white, rather like Tom. Smudge, the oldest female, had much brown in her fur and her markings were not so distinctively defined. She was almost tortoiseshell.

Tom made amatory advances to all three females. In late February and early March, though, it was Pickle who was receptive, not to say seductive. She greeted Tom with little chirping calls that grew into high-pitched 'wowl' cries. She rolled on the ground and rubbed herself against him. Sometimes Tom was seduced, but the strategy did not always succeed, and he would growl and rebuff her with his paws.

Life in the barn was usually peaceful and Tom's preference for

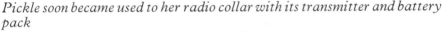

Pickle soon became used to her radio collar with its transmitter and battery pack

one female did not make him aggressive toward either of the others. When two cats met they often groomed one another. It was as though each animal had around it a space, enclosed within an invisible barrier, inside which it enjoyed complete privacy. Invasion of that space by another animal amounted to a violation that might be repelled. Entry was by invitation only, and a strange animal that invades the privacy of another is likely to be up to no good. After all, the barrier must be crossed if an attack is to be pressed home. Thus an invited infringement was safe, but the rebuff caused by an uninvited one was potentially so serious that the offending animal would withdraw at once.

Invitations could be solicited, and one way of doing this was to approach the other cat with gestures that were appeasing and unhurried, and to commence grooming. When one cat approached another without grooming, there was an even chance that it would be repulsed. If it groomed, the approach seldom failed.

Not all contacts were so demonstrative. Sometimes two cats would be content just to be close to one another, relaxed but apparently indifferent. At other times, the invitation would be to play. Now and then there would be a disagreement, a small display of aggression that ended as quickly as it began when one of the antagonists withdrew while tempers cooled. In most cases, the aggression was caused by an approach that perhaps was meant to be affectionate, but that proved unwelcome.

Peter recorded it all, the comings and goings, the everyday happenings, but as Tom settled down to rest in the hay, and as Peter began to eat his bacon and fried bread in his hut down in the rhododendron patch, the project was only just beginning.

The cats had been released to the farm rather earlier than had been planned. They had made themselves at home and clearly intended to stay. It was the weather that supplied the first test.

The 1977–78 winter, like most winters, was the worst in living memory. In Devon, it brought deep snow to a region of Britain unaccustomed to more than an occasional January flurry. High on the moors, houses were buried to their bedroom windows. Roads were marked only by the telephone wires running along at the new ground level, high above the vanished hedge tops.

Peter decided he was warmer among the hay in the barn than he was in his hut. The farm was cut off from the outside world and it would be some time before David arrived with the equipment he needed for the study. On most days he visited Maurice and Edith at the house, but there was one day on which he was not seen at all. It snowed for most of the day, and until well into the evening Peter sat huddled amid the hay bales watching the cats. He was pleased that they had settled in so well. The project was taking shape already, and Peter's mind was full of exciting questions which he hoped could be answered by his research. It was quite dark when he left the barn and found his way through the drifting snow back to his hut. As he was preparing to go to bed, the lights flickered and then went out. He thought little of it, assuming there was a total power failure as lines had been brought down by the weight of the snow.

The following morning the cause of the power failure was very apparent. The power cable to the hut ran from the barn, and the barn had collapsed! Maurice later estimated that the weight of snow on the roof must have been at least ten tons. To allow more light into the barn to make filming easier, part of the corrugated metal had been removed from the roof and replaced by transparent plastic sheeting. This had not been designed to support such a weight. As they stood glumly in the fresh, white snow, surveying the ruin, Maurice and Peter took it for granted that the cats would be dead or gone, for what sensible animal having survived so appalling a catastrophe would wait about on the off-chance of a repeat performance?

They began to pull and lift and lever, and little by little the wreckage was shifted. Not for the first time, Maurice was impressed by Peter's physical strength. He could pick up a hay bale and throw it to the top of a high rick. He needed all his strength as they sorted out what was left of Maurice's barn – the centrepiece of the whole project. It seemed that all their ambitious plans were in ruins. Then they heard the unmistakable cry of a cat. They scrambled over the bales and there, snug among the hay, were all four cats! If you are a cat, perhaps anything is preferable to wet snow, even a roof that caves in on you. It was clear, though, that offered an easy escape route the cats had chosen to remain. They were settled in, and from now on they would come and go from the barn whenever they pleased.

For the moment, though, there remained the problem of the barn itself, and the snow, which had begun to fall again. Mainly by brute force they rebuilt the barn as a much more robust structure. It took them three days. The snow continued to fall for some time, and in succeeding winters the elements were no more charitable, but the roof has not fallen again from that day to this.

Until they could fend for themselves, Peter had to feed the cats.

This had to be done subtly, for the cats must not learn to associate him with food. At the same time, they must not fear him, either.

At first he left the food – usually proprietary cat food – in strategic places where they would find it, but he made sure they did not see him with it. The cats remained hidden and cautious. By February he was leaving food once or twice a day and was observing the cats for four hours each evening and intermittently during the day. Maurice would wander into the barn sometimes to find Peter, his long legs sticking out from behind a hay bale, muffled against the intense cold, stolidly making notes. At that stage there was little enough to record, for only Pickle would emerge from the hay while he was there. He had yet to win their confidence.

Later in the spring, when they were beginning to explore the farm, he altered their food supply to a milk substitute called 'Denkavit'. By tradition, farm cats receive a dish of milk a day from the dairy. It bribes them not to migrate from the farm, but it leaves them hungry enough to make sure they hunt. This was the position toward which Peter was pushing them, and eventually, when he was sure they were catching food satisfactorily, he would reduce the supply still further. The ideal ration was calculated as 50g for each cat every day. In terms of nourishment, it was equivalent to about two-and-a-half mice. He fed them morning and evening, from two bowls. This made sure there was enough room for all four to feed and it guaranteed that Domino received her share. When he had used one bowl she, as the youngest of the quartet, had been pushed aside.

Solid food was given to them differently. Peter supplied it on a regular basis only during the early part of the project. He divided it into four portions, one for each cat, and placed the portions at intervals so there would be no quarrelling. It was the solid food that attracted Shadow.

Shadow was an immature male who seemed to have no home to go to and who lived by his wits on the fringe of Maurice's farm. He may have come from the village. Whoever he was, his behaviour suggested that he wished to join the group. The other cats gave him no encouragement, but Shadow proved to be a persistent visitor to the colony.

There were other cats in the neighbourhood. The Tibbles own a cat. Mizzy by name and ginger in colour, it is a large, neutered tom that had spent some of its time in the barn before the arrival of the new cats. On one of the first occasions that he fed the cats, Peter left briefly and returned to find Mizzy inside the pen and bloated, having consumed the entire ration for all four cats. After that episode, Mizzy retired to the house, conceding his rights in the barn to rivals by whom he was out-numbered and outclassed.

A tiny, disused, overgrown lane, said to be the oldest stretch of road in North Devon, runs from the church, past the lych-gate, and between tall hedges to the vicarage in the hamlet. Years ago, when the church had a priest to itself rather than sharing one with nearby parishes, the vicar would drive along the lane in his trap. The vicarage is now the home of the veterinary surgeon and his family, and they have cats. Tom visited them from time to time, mainly to steal food from them, and from there he would cut across a few fields, run along some hedge bottoms, and so reach another farm on which lived a solitary female cat with a deformed ear. This did not put off Tom, who soon made a habit of visiting her. Peter gave her the name 'Broken Ear'. In all, Tom was visiting five farms in the neighbourhood.

While Peter had been establishing these initial patterns of behaviour, David had been back in Oxford arranging the necessary technical support. Each cat was to be fitted with a radio collar consisting of a radio transmitter powered by a small battery. Each would transmit a different signal by which the individual cat could be tracked and its movements plotted. The larger the power pack, the

David and Peter fit Tom's radio collar

longer the life of the signal and the greater the distance over which it would transmit. The scientists preferred larger packs, but the film-makers thought that they would be visually obtrusive, and proposed small ones. It was decided to compromise and use packs that were intermediate in size. The choice proved satisfactory. Tom did manage to travel beyond his radio range once, but the packs were still transmitting up to the time when they were removed.

Initially, though, the collars had to be tried for size, using dummy packs, before the design was finalised. On 2 March Peter received two such dummy packs from David. He managed to catch Tom, only to find the strap was too short. Pickle gave him a bad time. He caught her, but her reaction to being handled was quite wild. Tom and Domino would allow themselves to be handled, the others would not, and that is the extent to which the four cats were 'tame'.

During the attempts to fit a dummy collar to Pickle, her screams attracted Tom and Smudge, who moved in to investigate. Peter had the distinct impression that they had come to Pickle's aid. What did this mean? Many cat enthusiasts who are also cat watchers (it is not necessarily the same thing) believe that cats are social animals, ready to assist their friends, but the scientist is permitted no such simple picture. If the cat is a truly solitary animal, this does not mean that it must scorn entirely the company of its fellows. Mothers must be allowed to bring up kittens, of course, but larger groups might assemble for particular reasons. They might meet to exploit a plentiful source of food, for example, or to shelter from bad weather. Such contacts might well be amicable. There is no reason to suppose that individuals will quarrel without cause, and a solitary animal might welcome the comfort that is to be derived from mutual grooming and from sleeping huddled against another cat's warm fur. Such loose aggregations of cats have been seen many times and by almost everyone. If we assume that the cat is a truly solitary animal, always and inevitably, we go on to assume that such groupings of cats can be no more than loose aggregations. The individuals that comprise them each live their own lives in their own way, pursuing their own self-interest.

It is a view of the cat that will appeal to many people, for it confirms their prejudices. It reveals the cat as a robust individualist that goes its own way and does exactly as it pleases. If it adjusts its behaviour to suit the convenience of the humans whose accommodation and food it shares, this is a reasonable pursuit of the cat's own best interest, since persistently anti-human behaviour will bring ejection on the toe of a

Pickle and Smudge at home in the barn

boot. Looked at another way, the cat is exposed as a shamelessly self-centred, greedy, hedonistic individual, and because it must be credited with a certain low cunning in the pursuit of its nefarious ends, it is deceitful, unscrupulous and dishonest. The solitary cat is a very satisfactory fellow, for he fits neatly into the pigeonhole we have prepared for him mentally, and allows himself to be loved or hated, admired or despised, according to taste. That the cat might be anything but solitary by nature seems uncharacteristic.

What if the picture is quite wrong? What if circumstances can arise under which groups of cats will live together not as a loose aggregation, but as a society, a group whose members collaborate now and then in pursuit of a common objective, or who even behave altruistically toward one another? Was the reaction of Tom and Smudge to Pickle's distress during the collar incident simply curiosity, or was it an act of altruism? As the cats settled in, would they show that they had a social side to their nature and way of life?

By the third week in March the cats were roaming free. They had started to hunt. Peter saw Domino catch, play with, kill and eat a vole. They were beginning to hide food in caches for future attention. Tom even tried to bury a dish of milk. Peter was confident that the cats were familiar with their surroundings and were starting to fend for themselves.

Domino had become sexually receptive. She was the smallest of the cats, and it is possible that she was a virgin. When one of the other cats touched her, she would crouch and rub her chin on the ground. Tom would bite her neck in the approved manner, but her behaviour when he mounted her suggested she had no previous sexual experience; and it took several attempts before she learned to tuck her legs beneath her. After about a week she was responding to seductive chirps from Tom by crouching and rolling on the ground, but when he tried to grab her, she moved away. Domino's confused behaviour was a sign that she was now sexually mature. Soon she would be physically responsive to Tom's courtship.

Peter finished his breakfast and returned to the barn. The cats, asleep or pretending to be, ignored him. A morning hen clucked hysterically to announce she had just invented eggs, and the old buzzard lifted himself into the warming air to ride the thermals while he waited for the flicker of movement that would allow him to break his fast. It was morning. It was spring. For Peter it was a time to wait and watch.

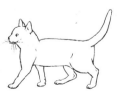

Staking out territories

As Tom walked into the barn, almost the first thing he saw was Shadow, sitting on top of a hay bale. The three females were taking little notice of him, but Tom was affronted. He was prepared to tolerate Shadow, but liberties he would not permit.

He jumped on to the hay as though to attack. Shadow crouched down and hissed loudly. It was impossible to tell whether his attitude was one of defence or submissiveness, but it was clear that he recognised he was in trouble. The two animals stared hard at one another for about five seconds, while Tom remained perfectly still and in control of the situation. Then he began to step from foot to foot with his front legs splayed outward and forward, claws fully extended, tearing at the hay. Still he made no sound, but vocal threats were hardly necessary. The menace in his regular treading and tearing, treading and tearing, was plain, and it was impressive. Then, abruptly, he stopped, turned away, and twitched his tail and hind quarters in an action that suggested to Peter, who was sitting very still a few yards away, that he might be scent-marking. Supremely confident, Tom sat down with his back to Shadow.

He remained like this for about a minute, but it seemed that his message, clear though it had been, had not succeeded in dislodging his opponent. He turned and repeated it. This time Shadow retreated about six feet, hissing and growling. Tom could probably have taken him then, with a leap that would have found Shadow in a weak position for defence. He did not press his advantage, however, but turned his back and sat down again. The stalemate continued.

The cats remained immobile and about six feet apart for a further half hour. They appeared relaxed, but there was a tension between them, a conflict that had not been resolved. Suddenly Shadow moved toward Tom, startling him into jumping away, but still there was no physical contact between them. The war was purely psychological, conducted through threat and counter-threat.

For another twenty-five minutes the cats sat six feet apart without moving. Then Tom, perhaps deciding that enough was enough, walked away, his dignity preserved and, it would seem, his honour satisfied.

It was time to feed the cats. Peter stood up and began to move about the barn. At this, Shadow ran outside and vanished.

The argument between Tom and Shadow was probably territorial. Shadow had invaded a region Tom considered his exclusive preserve. There were other such encounters, with other cats, but these took place in various other places about the farm. Only Shadow invaded the barn itself.

There was a ginger cat – not Mizzy – who caused Tom trouble. Maurice once saw Tom fighting him in the churchyard on the far side of the house from the barn, and a little while later Peter watched Tom chase a ginger cat – perhaps a different one – through the orchard and across the road. It was only Tom who was so aggressive, though. The females never went out of their way to find trouble, as he seemed to do, but their attitude to Shadow was curious. He was the one cat the females would attack. In this their behaviour was not typical of their behaviour to other cats. Pickle, especially, seemed to resent his presence. On one occasion she crouched behind a hay bale in order to pounce on him as he passed.

Tom behaved inconsistently, as well. Sometimes he would try to attack Shadow fiercely; at other times he took no notice of him. His attacks never led to physical contact, but he was not content to allow Shadow merely to run away, as he allowed other intruders to do. He would cut off Shadow's retreat, forcing a confrontation, but then he would allow Shadow to escape unharmed. Shadow himself was always nervous. He appeared only at feeding times and he always retreated, or tried to retreat, at the first sign of hostility.

The relationship between Tom and Shadow was more complex than it seemed. Had Tom simply hated the other cat and wished him gone, he could have pressed home one of his attacks or repeatedly chased Shadow away, in either case ensuring that Shadow departed once and for all. That he never went so far, suggested that he was not seeking to exclude Shadow totally, but to make sure that his visits were restricted and that he remained subordinate. This, too, would account for his apparent inconsistency, for it has been observed in other studies of cats' behaviour that provided the enviroment imposes no extreme pressures, a group will be dominated by a male but the 'boss cat' will not seek to assert himself constantly. He will react to a challenge from a subordinate, but otherwise his régime is usually placid and fairly liberal. He may not even intervene when another male copulates with one of 'his' females. Shadow was not a member of the group, however, but a worrying intruder who neither quite belonged nor quite did not belong. Tom had to make sure that Shadow understood the position, and his status required occasional affirmation.

This degree of toleration may well have been due to the fact that Shadow was immature at the time, a kind of feline adolescent. The threat he represented was potential rather than actual. Had a fully mature male arrived on the scene and sought to establish himself, he could have done so only by assuming the dominant role, and things would have been very different. Tom would have treated such a cat as a credible challenger and would have reacted accordingly. Almost certainly he would have won the encounter. The fact that he was fighting 'at home' would have made him more confident and the challenger more nervous.

All the same, as spring wore on and Tom's absences from the barn became longer, Shadow grew increasingly impertinent. He would flee from the females only if they were particularly aggressive. It was Tom he really feared, and when the 'boss' was away from home he ventured more boldly into the barn.

Clearly, Tom, and to a lesser extent the females also, were staking out some kind of territorial claims about the farm. They were defining boundaries. The local cat population was informed that this was where the four of them now lived and hunted, and that incursions would be rebuffed. Does this mean that cats are territorial animals?

To answer this we must distinguish between the terms 'territory' and 'home-range'. A territory is a region occupied by an animal or group of animals into which no outsider is permitted to enter. European robins are famously – or notoriously – territorial. This can be demonstrated very simply by placing a robin-sized object with a patch of rust-red colour on it anywhere inside a robin's territory. The resident robin will attack it. In the same way, it will attack a tape-recorder that plays a recording of a robin's cry, including a recording of its own cry. Robins, then, are aggressively territorial. Some monkeys, too, devote much time to occupying the boundaries of the group territory and screaming abuse and defiance at neighbouring groups. As with the robin, an invading individual will be attacked. Such monkeys are territorial.

Other animals behave in less obvious ways, and the fact that they have reputations for aggressiveness provides no clue to their attitudes toward other members of their own species. Weasels, for example, which will attack animals many times their own size and can inflict quite severe injuries on humans who attempt to handle them, are very tolerant of other weasels and fights are rare. Weasels have ranges, and they avoid border incidents by the simple expedient of avoiding direct contact with their neighbours. Should an individual stray into its neighbour's patch, no dire consequences follow. Weasels, then, have home-ranges rather than territories.

Territories and home-ranges serve similar purposes in that they are both areas within which an individual or group has exclusive or dominant feeding rights, but whereas a territory will be defended, a range may not be. It is marked, so that all animals know its boundaries more or less, but it exists by mutual tolerance and understanding rather than requiring maintenance by force.

The terms are not mutually exclusive. A species can defend a territory while simultaneously occupying a much larger range. The territory, forming a 'core area' within the range, is used for resting in safety and for breeding.

Whether an animal has a territory or a home-range depends, more than anything, on the area it requires to keep it supplied with food. As the cats began to locate the best hunting grounds and the most convenient routes to and from them, the area defined by their regular travels amounted to about half a square mile. It would have been impossible to defend so large an area. Even had the cats devoted the whole of their time to patrolling the borders systematically, there were too few of them to have provided more than the most sketchy protection. In fact, it was only Tom who patrolled the outlying parts of the ranges, because the females had much smaller home-ranges which, for the most part, were totally contained within the patch he had staked out for himself.

Should Tom encounter another male inside his borders, he would order it to leave, which invariably it would do, but should a male – or female for that matter – enter without his knowledge, no harm was done unless the intruder stole food from him or attempted to mate with one of his females. What he could not permit was the establishment of a resident competitor. Peter and David assumed that Tom's range was based on the minimum area he needed to keep him fed. The fact that his boundaries lay beyond those of the females in his own group helped them, for it meant that not only were they less likely to encounter intruders, but they did not have to compete for food with Tom, either. He and they spent much of their hunting time at or close to the

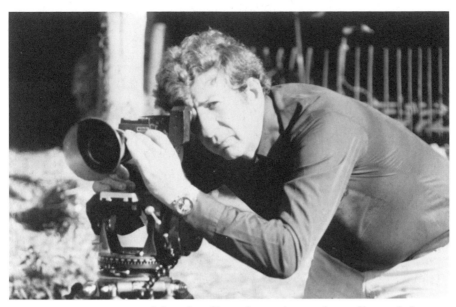

Maurice films the curious private lives of the cats

boundaries, and quite quickly their movements began to follow regular paths. They had 'beats' that they walked, probably less for fear of invasion than to inspect their favourite feeding spots.

Maurice used to see Tom trotting past the house on his way to or from the barn. He walked part of the way beside the hedge that bounded the Tibbles' garden, but then, and always in the same place, he went through it to the field beyond. He approached the house through the churchyard, entering by the lych-gate and following the path that led beside the graves to the door. On at least one occasion Maurice saw him scent-mark the church door with urine. Tom made enemies as well as friends!

From the church his regular route took him north-west to a farm, then almost due west for about a quarter of a mile, then south, parallel to the road that runs up to the church, but to one side of it by a distance varying up to about 200 yards, and past two more farms, where he would sometimes call to inspect. The fact that the farms lay within his range, and that they harboured cats, meant that those cats were, or should be, subservient to him. Such was Tom's simple political philosophy, demonstrated now and then by a scrap that, over the years he had previously lived in Oxfordshire, had tattered his ears and given him his practical, no-nonsense attitude to life. So he continued, a one-cat occupying army, along a track that led him almost a mile to a field boundary where he turned left to walk due east and back to the road. He was heading now past some cottages and towards another farm, where lived Broken Ear. As Tom approached her barn, he would raise his tail in greeting. It was apparent to Peter, who pursued him, that Tom was well known here and that he was made welcome by the lone female.

After leaving Broken Ear, Tom would head north-east for just over a quarter of a mile until he was east of the village. He skirted the village to the north – although he would often make a diversion into the village – and returned to the church by the side of the ancient lane along which the vicars used to ride.

Behind him plodded Peter, the hood of his parka raised and his antennae held high above his head. He wore headphones and twiddled intermittently with the controls of the radio receiver that hung by a strap around his neck. Sometimes though, Tom would wander for days on end, disappearing beyond radio range. Peter recorded:

Tom continues to roam the countryside. I did not see him between the evening of 8.5.78 and the morning of 11.5.78. He was also away between the evening of 12.5.78, when I tracked him to the village, and the morning of 14.5.78. On 13.5.78, he was out of radio range along all three sides of a triangular area with about one mile sides. The tracking results so far suggest that he is fitting the expected pattern of restricted hunting areas joined by well defined paths.

A few days later:

Tom continued to leave for the village at about 7 o'clock until 18.5.78. On 19.5.78 I intercepted him on his way back at the village end at 0840. He arrived back at the barn at 0910. On the night of 19.5.78 he did not leave and was still in the barn at 0400 on 20.5.78. He did not leave on 20 May and is still around tonight. This change of habit unfortunately coincided with the need to watch Domino intensively, so that I did not track him on 18.5.78. Something may have happened at the village which frightened him off. Possibly coincidental was the sighting of a large, pale ginger cat near the church on the evening of 19.5.78. In the morning there was an area of flattened plants and some tufts of pale ginger hair to the east of the churchyard – presumably the site of a cat fight, not, presumably, involving Tom, since he was in the village.

Like any sensible cat, Tom was an opportunist. He was a good hunter, but in his view there was no point whatever in spending hours waiting for a mouse when humans provided food for the taking. There was a villainous purpose for his frequent visits to the village.

The vet's wife owned two pet cats who were fed regularly on good, chewy meat, which they nibbled delicately, eating a little now and saving a little for later. They were no match for Tom. Their owner became distressed at what seemed to be an abrupt change in their habits. They would be fed in the usual way, but a few minutes later they would be demanding food again, and their plates would be empty. Such gluttony was not to be encouraged, and the cats were told so, in no uncertain terms. This happened on several occasions, and when the owner glanced through the window, she saw Tom looking complacent and licking his lips, but thought little of this. It took some time for her to realise what was happening, and when she did, Tom made another human enemy!

If you are a cat that lives by its wits, the secret of success lies in rapid eating. Like most such animals, Tom did not stand on ceremony or toy with the food. He ate like a vacuum cleaner, cleaning a plate in seconds, a second plate in a few seconds more, and he was through the door and away in less time than it takes to tell. No doubt this technique caused appalling indigestion, but it was easier than hunting, and probably just as enjoyable.

He raided Maurice's house, too, when the coast was clear, and his tastes were catholic. He once devoured a chocolate cake that Edith had baked, and the experience left him with a taste for chocolate cake that he never lost.

Peter spent part of his time working as Tom's public relations officer. It was a thankless task that took him to local farms and houses, where he had to explain and apologise for the behaviour of a cat who was having the time of his life. As the subject of scientific research, Tom could not be constrained in any way. He had to be allowed to do as he liked, for the purpose of the study was to observe just what he did like. And if that were not justification enough, by this time Tom had been cast as a star of the small screen, and film stars have to be humoured.

But there were occasions when Tom's extra-mural activities had to be restricted.

Maurice's farm also provides a natural setting in which he carries out some of his work as a wildlife photographer, and that season the cats were not the only animals around the place. Maurice had a family of water-voles living near to the house. The female had just given birth to a litter in front of the camera, an event that had never before been recorded on film, and she was beginning to suckle her young. He was making a film about the life of the water-vole for the BBC series *Wildlife on One*, and everything was going well until Scott, his son, reported seeing one of the cats uncomfortably close to the water-vole nest. It was too late. One of the cats – it may not have been Tom this time, but he was the prime suspect – had eaten all the youngsters. The vole enclosure was then made cat-proof.

Tom's activities, and those of the females whose ranges overlapped with one another, illustrated the purpose of the home-range itself. Each animal located a series of feeding places and spent its time travelling by regular paths to each of them in turn. It was a hunting operation that had nothing to do with defence. In Tom's case, though, his travels included visits to neighbouring females. The boundaries of the ranges were clearly defined, and they represented the furthest distance each animal could travel from 'home' – the central area where the females slept, with or without Tom – without feeling insecure.

The females were more social than Tom, spending much more of

Domino exploring the meadow beyond the barn

their time together, but when Tom was present they behaved differently. He protected them from other cats, and they were noticeably less nervous when he was with them.

When members of the group met one another away from the barn, as the females did now and then in the course of their hunting, they greeted one another by sniffing, rubbing heads and flanks against one another. Sometimes one sniffed the tail region of the other. Very early on in their exploration of the farm, it was clear that they remembered places where they could find food and they had no trouble returning to them, navigating by sight but also, and sometimes more importantly, by following a trail of smells.

As their confidence increased, so did the size of their ranges. By the end of June, Peter reported that Tom had doubled his range area and that Smudge and Domino had increased theirs by half. Sometimes Tom ranged further, but his progress to the north was checked one day by a bullock, which chased him into a wood. He continued to meet other cats.

On the night of the 5th (June), he took part in a caterwauling session for about 15 minutes at 0100 with a large dark cat. As far as I could see (not very far) they just sat about six feet apart and wailed at each other. The other cat finally walked off and Tom sat nearby for an hour until I left him.

. . . While at the village on 10 June Tom met a large (bigger than Tom) black and white cat. Although they did not fight, the black and white cat twice ran towards Tom and either sat about fifteen yards from him or ran past at about nine feet distance. After one such encounter, Tom spray-marked a nearby gate post.

Tom was spray-marking widely. Apart from the church porch, he had sprayed many times in the barn itself, on gate-posts, and in the small woodland just down from the church, which developed a distinctly 'catty' smell.

Males spray much more frequently than females, but the activity is not exclusively male. Domino sprayed at least once, going through the full routine and obtaining the appropriate reaction from Tom, Smudge and Pickle, all of whom were present at the time.

Why do cats do this? The conventional explanation is that spraying supplies marks by which territories are defined, and that a series of marks on prominent surfaces warns other cats that they are entering occupied territory. As with so many aspects of cat behaviour, this is not an adequate explanation. Why, for example, should the cats spray inside the barn, since the recipients of such 'messages' would be principally themselves?

In the first place, what is it that is sprayed? Tigers, lions, cheetahs, in fact all members of the cat family, including the domestic cat, spray, and all of them do it in the same way. It has been suggested that tigers and lions are able to mix some pungent secretion from glands around the anus or genital region with urine, but whether or not this is so, and whether domestic cats also add some special ingredient to their urine, no one knows. All we can be certain of is that the spray consists mainly of urine. A cat urinates in a downward direction – on to the floor – but it sprays more or less horizontally and is unable to spray downwards. Some cat observers have found that cats will react to the smell of urine whether they find it on the floor or at a higher level. This suggests, therefore, that in the domestic cat at least the substance that is sprayed is urine and nothing else. Peter's findings, however, tended to contradict this. David sent Peter a supply of cat urine and Peter used it to make scent-marks which he presented to Tom. At first Tom took no notice whatever. This may have been due to the age of the urine; fresh urine might have aroused more interest. However, Peter also saw Tom come across some ordinary, unsprayed urine left by Shadow, and again Tom took no interest.

Cats do possess scent glands, which are located on the side of the head and in the angle of the jaw. They are used for marking, the secretion being transferred whenever the cat rubs its face and head against an object – the object often being animate. Every cat owner is familiar with this, and when Pussy rubs herself against our legs or hands we like to imagine she is making a simple gesture of affection. Not so. She is marking us with a scent she will recognise the next time she encounters it. It is her way of providing 'badges' to distinguish between friends and strangers quickly.

A cat may rub itself against the target it intends to spray, but it does not always do so. The spraying itself always follows the same ritual. The cat turns its back on the target, lifts its tail, and in a standing position it sends a stream of urine in a horizontal, or nearly horizontal, line. The amount of urine involved is much less than the amount excreted in ordinary urination. During spraying the cat's tail quivers and it may make treading movements with its hind legs. When spraying is finished, the cat may sniff at the target, but if it does so at all, the sniffing is very cursory. Normally, cats bury their urine and faeces: there is no advantage to a predator in announcing its presence to its prey or to larger predators. Sprayed urine is never buried; it is left for all to find. The urine is meant to smell, and that is why any sniffing of it is cursory. But the division between ordinary urination and spraying is not hard and fast, and sometimes it is difficult to tell which action a cat is performing.

There is a general rule among cats that fresh spray-marks, usually those that are less than four hours old, must not be sprayed again, but older marks are renewed often. An individual male may spray in one place only or in several places, but once chosen, a particular site is likely to remain in use for some time.

This makes it possible to 'translate' the chemical message into a verbal one. Paul Leyhausen, a German zoologist who has devoted years to the study of cat behaviour, suggested in 1971 that the spray-mark simply means that another cat has passed this way. If the mark is fresh, the cat has passed recently. If the mark is old, then the cat that finds it may proceed, but before doing so it should mark the spot again so that the next cat will know what it is expected to do. In other words, the marks are intended to avoid confrontations between cats whose ranges overlap. At least, that is one of their suggested purposes. They may also have sexual overtones. This can be inferred from the behaviour of a cat when it discovers a scent-mark.

This behaviour is unmistakable. A cat explores its environment by sniffing, and smells are at least as important to it as sights and sounds. Most of this sniffing is fairly casual, but when a cat comes across a scent-mark, the sniffing becomes more intensive. Its nose actually touches the marked spot, its ears are laid back, and it may begin to make licking movements with its tongue across the roof of its mouth. After this, an activity that students of animal behaviour term 'Flehmen' may occur. The cat raises its head with its mouth half-open and its upper lip drawn back a little. It stands very still, stares fixedly in front of it, and breathes slowly. Then the cat may sniff again and Flehmen may occur for a second, or even a third time.

Peter described the process as part of the spray-marking experiments he conducted in June.

When a mark was dabbed on to the upright post at the front of the barn, Tom picked up the scent from about a yard, sniffed carefully over the post and found the mark. He sniffed very carefully at it, touching it occasionally with the tip of his tongue, and showed Flehmen. Tom spray-marked four spots around the barn during the morning. The females took no notice of them.

Flehmen is not restricted to domestic cats. Some of the large cats exhibit it during sexual intercourse, and male cheetahs perform it more frequently when they are sexually active. Domestic tomcats are likely to perform Flehmen more often if the urine they are examining was left by a female in oestrus. It is such clues as these that suggest scent-marking has some sexual significance, but it is feral goats, not cats at all, that demonstrate such a connnection most dramatically. When a male goat comes across a scent-mark left by a female that is not in oestrus, it performs Flehmen and after that its sexual behaviour is inhibited.

On 7th April, Tom had broken all the rules. After adopting an unusual stance to produce it, he left faeces unburied – quite deliberately. Was this some novel form of scent-marking? Peter, who observed the event, could not at first think of a satisfactory explanation, but it was not the only unusual behaviour he saw. Earlier on the same day, Smudge and Tom were together in the field in front of the barn. Smudge rolled, crouched and rubbed herself on the ground. Tom chirped seductively and tried to bite her neck. Twice she pushed him away, but only to repeat behaviour that is typical of a sexually receptive female. Tom mounted her repeatedly and tried to copulate, but did not succeed. Smudge looked pregnant, but had she been pregnant she would not have been sexually receptive. On the other hand, Tom's failure to copulate with her made Peter doubt whether she was really receptive at all.

The following day the cats were all sleeping together when Pickle woke and began nuzzling at Tom. She rolled him over and nuzzled his tummy, with her eyes closed and her ears laid back, and after a couple of minutes she found a nipple, which she sucked for about 30 seconds. Then she released it and nuzzled again, kneading with her forepaws. Tom purred loudly throughout the proceedings, and when Pickle sat back he rolled to his original position. 'Any possible functional significance eludes me!' noted Peter.

The cats were living together contentedly. In the early days, Domino had often behaved submissively to the others. While Peter was supplying all their food, during the settling-in period, he had to take special precautions to make sure that Domino received her proper ration. She would be pushed out of the way if she tried to feed with any of the others. Now, though, she was becoming more popular. Pickle was generally friendlier, especially with Domino and Tom, and Smudge was getting along better with Domino. This change in relationships, which Peter described as 'sharp', seemed to end any social ranking. There was no open dominance of cat by cat.

Domino and Tom, mutually grooming

Indeed, the peace that descended in the barn was starting to bore Peter, who noted that: 'the discontinuous behaviour records seem to reveal little except that cats spend a lot of time doing not much!' and, later, that the record showed 'the expected large amount of sitting, sleeping and grooming'. When a cat returned to the barn after spending some time away, it was often greeted by the others with shows of affection.

The cats were usually in the barn in the morning. As spring moved gradually towards the slow, sleepy, North Devon summer, they would sunbathe together. The afternoons and evenings were the time for exploration and hunting.

Exploration was extremely important. Far from killing the cat, its curiosity about everything around it is part of the secret of its success. This curiosity may seem playful, but it is neither trivial nor idle. Every possible source of food must be discovered and tested. Every possible source of danger must be located.

Once each animal had explored its range to its own satisfaction, life settled down to a more regular pattern. Tom, with his large range, spent his foraging time walking briskly, sometimes trotting, from one place to the next. He spent most of his time on the move, seldom

Domino returning from a successful hunting trip

staying for long in any one place. The females were less active. Smudge, for example, would visit the farm halfway down the road during the daytime. Peter found it difficult to track her among the farm buildings, but she appeared to spend long periods in one place. At night she moved into the open countryside. It was the females who sat patiently by the mouths of burrows waiting for their occupants to emerge. Tom, of course, was still living largely by larceny.

The difference in range sizes between male and female was apparently typical. In her study of cats living in Portsmouth Dockyard, Jane Dards found that on average toms have ranges larger than those of females and use their ranges in a more uniform way. Three-quarters of the females she observed lived in groups, sharing ranges among two or more adults. The actual size of a range seemed to depend on the resources it contained, but was independent of the number of cats in a group. The females tended to spend most of their time in a small core area, where they had food and shelter. The frequency with which they moved away from this area varied from one cat to another and seemed to depend on individual temperament, some cats being naturally more adventurous than others. The toms spent their time, as Tom did on the farm, moving from one site to another within their ranges, but many of their visits were to the core areas occupied by groups of females. It is the toms that wander far from their usual haunts, most commonly in January and February, when female cats start to come into oestrus.

It is not difficult to see how these patterns of behaviour typical of males and females benefit cats as a whole. Females have to bring up kittens and for quite long periods their movements will be restricted because very young kittens cannot be left unguarded, or unfed, for very long, and slightly older kittens cannot travel far. The females, therefore, find it advantageous to settle in a secure place that is surrounded by a fairly small, but rich, range in which plenty of food can be found. Toms suffer no such constraints. Their contribution to the continuation of the species requires them to mate with as many females as they can find. An unattached receptive female is likely to mate with the first acceptable tom to arrive on the scene.

It is all managed very efficiently, especially if the use of ranges is interpreted in terms of its genetic benefits, since it ensures that the genes possessed by each individual have the largest possible chance of surviving. The size of the female ranges permit their occupants to provide care for the young, so preserving the genes shared among the family group until each new generation matures; the male ranges tend to introduce those genes into new family groups where they will be protected. By visiting several family groups, each male provides 'insurance' for the survival of the genes he carries against the possible extinction of any particular group.

By the third week in May, Tom was becoming increasingly irritable. He was inconsistent, changing from being aggressive to being friendly in a matter of minutes, but several times he struck Domino with his paw and she became very cautious of approaching him. Peter thought this was due to hunger. The four cats would all drink together from the same bowl, but if they were given solid food they would growl at one another. If the food consisted of prey one of them had caught the aggression was more marked. On one occasion Tom ate a mouse, growling all the while at any cat that came within a yard of him, but as soon as his meal was finished he groomed Smudge's back.

The cats were hunting, and like all farm cats, they were now largely dependent on the food that they individually caught. In this respect, it was every cat for itself.

Hunting

Assination

CATS, OF COURSE, are formidable hunters. At least, they are if they have been taught to hunt while they were kittens and have kept in practice. Many pet cats never catch anything, and although they may go through almost ritualistic stalkings, often of somewhat inappropriate quarries, these rarely develop into full-scale attacks, and when they do the cat frequently misses. The desire to hunt may be innate, but the skills of the successful hunter are learned and perfected by practice.

The successful cat works in classic fashion. It sits patiently beside a hole, waits for the occupant to emerge, then pounces before the prey has a chance to react. All our cats spent some of their time sitting beside holes, and Tom particularly resented any intrusion when concentrating on such important activity. When Domino approached him with her usual greeting she was rebuffed, and she soon came to realise that play may be play, but work is work, and the condition of the inner cat imposes its own high priority.

Hunting by ambush is characteristic of all the cats, large and small, with the exception of the cheetah, which hunts on the open grasslands where the strategy would not work so well. Even the cheetah, though, cannot run its victim down as a dog might do. Cats are designed to move very rapidly, but only over short distances. Their stamina is poor and they cannot sustain a chase.

Nor, curiously enough, are they very good at climbing. At least, they can climb up well enough; it is descending that causes problems. Everyone knows that a cat can fall a considerable distance, land on its feet, and walk away physically unharmed. The cat up a tree seldom uses this remarkable ability. It runs up a tree almost as quickly as it moves on level ground, using its extended front claws to grip the bark and its strong hind-leg muscles to push it forward against the anchorage provided by its rather immobile back claws. Unhappily, the technique does not work in reverse. There are tree-climbing wild cats, such as the American ocelot, that are able to descend trees head-first, moving spirally down the trunk. They have hind feet that are very flexible and all their paws can be used for gripping. Were the domestic cat to try such a vertical descent head-first, it would be unable to maintain a grip with its hind feet, because the feet cannot be turned around and the claws are arranged so that they grip only when the legs are thrust backward – they curve over and down, and were they to continue, they would point to the rear. The front claws could grip the bark much better, but the sad result of this would be that the cat would be secure at only one end – the front end. Its hind claws would part company with the tree and the rear part of its body would fall away, causing the cat to tumble tail over head. The cat is well aware of this danger. Its solution is to descend tail-first, using its hind claws to grip as they are meant to grip, holding the weight of its body, while it releases its front claws a paw at a time and brings them down. The operation is clumsy and slow, and while it is descending it is difficult for the cat to see where it is going. The whole thing is most unsatisfactory, and although many domestic cats learn to conquer their understandable fear of it, many never do. The cat stuck up a tree elicits our sympathy, and fire brigades are fairly expert at cat rescues.

It is not unknown for a cat to exploit this. After seeing 'The Curious Cat', one *World About Us* viewer wrote describing how her pet cat would climb a tree and howl miserably for help, so attracting a friendly, attentive crowd, but if the crowd dispersed without effecting a rescue, the cat would descend without the slightest difficulty and try again later.

The domestic cat is designed to hunt on the ground, and if its claws and muscles are poorly constructed for climbing, they are superb tools for catching and despatching small prey.

The hind claws cannot be moved, except as part of the digits to which they are attached. The front claws are normally held in a retracted position but can be extended when the cat wishes to use them. The claw is attached to the outermost of three small bones in the toe, but two tendons, above and below, attach it to the bone behind. When the upper tendon is contracted, the outer bone and claw are pulled up and back into a sheath of skin in the paw. When the lower tendon is contracted, the bone is pulled from a position alongside the middle bone, so bringing the claw forward and down into the extended position. All the cats, large and small, can extend their claws in this way, with the solitary exception of the cheetah, whose claws are extended permanently.

The normal position for the claw is sheathed. Inside its sheath, it

Smudge devours a mouse – head first, in classic cat fashion

is protected from the wear, tear and blunting from which the thicker and tougher hind claws suffer. The motion that extends the claws is accompanied by other muscular contractions that spread the toes wide, so that the paw attains its maximum span. Left unattended, both front and hind claws would soon be covered by the protective sheath of skin. Apparently neurotic cats that sit by the fire 'biting their nails' are in fact biting away the claw sheath. Front claws are kept sharp and the outer, horny layer is removed from their tips by honing them on any suitably tough surface. Carpets and furniture are excellent for this purpose, but for cats that live outdoors, trees must be made to serve.

As anyone will know whose games with a cat have become rougher than either combatant planned, the hind claws are useful weapons in their own right. In desperate straits a cat can roll on its back and fight upward, gripping with its front claws while it bites and at the same time rakes with its hind claws. It is the front claws, though, that grip.

That is their only use in hunting. They grip the prey, but they do not kill or injure it. Once it is held still, the ultimate weapons can be brought into action.

The killing bite of a cat is very precise indeed. In all members of the cat family, the death blow is a bite that severs the spinal column of the prey. The feat is more difficult than it sounds, because it is not achieved by crunching through a neck vertebra. This would be impossible in most cases and, even if it were possible, immediate death would not be the inevitable result. The sharp, very pointed canine teeth locate the joint between two vertebrae, and then they bite so that the bones are levered apart and the neural cord is cut. Death from such a bite is instantaneous, and the need to deliver a neck bite is so deeply ingrained that many cats, and especially the large cats, will often bite the neck of their prey ritualistically even when it has died from some other cause.

Everything depends, of course, on the cat being able to find the appropriate part of the neck and deliver its bite. There is no guarantee of this, especially if the prey is large compared with the cat and fights back. In such a case there are various possibilities open to the cat. The one most frequently adopted is for the cat to strike the victim with its paw so as to knock it off balance or, if it can manage to grab hold of the prey, actually to pick it up in its teeth and throw it violently to one side, perhaps repeatedly. The intention is to stun the victim, leaving it temporarily disoriented and immobile and so allowing the cat to attack the neck. Although the back of the neck is the most vulnerable point in any small mammal, it is not always the most vulnerable part of a bird, and killing a bird, especially a large bird such as a pigeon, can be difficult for a cat.

For a mammal larger than a vole or a mouse, turning to face the cat is often enough to stop an attack. The cat does not expect it and is confused for long enough to allow an escape. Tame rabbits, which have learned not to fear other animals of about their own size, are not really at risk from cats. Indeed, they often succeed in driving away a cat that troubles them. The method calls for strong nerves. The rabbit must graze in its usual way and appear not to notice the cat that stalks it, although it may move slowly and naturally, without looking around, to draw the cat into more open ground. It must allow the cat to get within striking distance. At this stage the cat is not committed to an attack, but has positioned itself to make one. Once it leaps, covering the last yard or two in a few large bounds, it is committed – and the rabbit will need to jump aside, which it is well able to do. As the cat prepares to leap, the rabbit must turn to face it, and then approach. The cat will be startled, then puzzled, and it will retreat to consider this curious turn of events. In this state of mind a further approach by the rabbit will make the cat retreat again, and if the rabbit advances every time the cat tries to advance, the cat will be thoroughly unnerved and will soon abandon any idea of attacking. Indeed, it may feel itself lucky to escape at all, and such confrontations sometimes end with the rabbit chasing the cat.

Were cats to collaborate in hunting, the rabbit would stand no chance, since it could not face two simultaneous attacks from different directions. Most of the cats, though, are solitary hunters. The lion is the famous exception. Like the domestic cat, it would prefer to sit and wait for prey to come to it, but on the African plains where it lives it cannot obtain animals in this way. Its victims are members of herds, and they are very watchful. Nor can it use the strategy of the cheetah, for it is not fast enough to overtake a long-legged zebra or antelope, even in a short chase. So it has evolved its own version of the ambush. A group of lions, usually females, moves into position, very carefully and behind cover, and one or two other members of the pride – often males – then appear to the prey, threatening them very obviously. This makes the victims run in a predetermined direction, and so into the trap.

Domestic cats do not hunt in this way. They have no need to do so, and for them it would hardly be practicable. Lions kill large animals and the food is shared among the members of the pride. There is not much of a mouse to share! There are plenty of mice, though, and plenty of good places in which to sit while waiting for them.

What the cat is prepared for is a small animal that moves by scuttling along like a vole, or in short bounds like a fieldmouse, and that has a clearly defined neck. If it has no obvious neck, the attack on it may be misjudged or even abandoned. If it has a neck, but is much smaller than the cat, turning to face its attacker may be of no avail. A wild rabbit will not stand fast to a cat, of course, but a rat will, and a full-grown rat is a formidable opponent. A dog can overwhelm it by its superior size, shaking it until its neck breaks, but for a cat, a rat is a much more even match, and not all domestic cats will overcome their fear and fight to kill.

Our farm cats had no such fears and, before long, rats formed a significant part of their diet. Many of the rats were young and fairly harmless, but now and then one of the cats would bring back an adult rat, and on at least one occasion Peter found a rat in the barn that had been killed by a neck bite but not eaten.

Cats eat their victims almost always from the head down, and experiments with them have shown that it is not the position of the head that matters, but the lie of the fur. Birds are eaten in the same way,

Smudge was the most skilful mouser

but sometimes a cat will start eating at a wing rather than the head. The principle is the same: the outer covering of the prey must go down the throat smoothly. Some people have speculated that just as fur and feathers evolved from reptilian scales, so this feeding behaviour evolved as a way of dealing with them. Perhaps this is so, but the immediate advantage of feeding in this way is plain. Were fur or feathers to be pushed into an erect position as they passed down the throat, the predator might well choke.

Sometimes the cats shared food, but not regularly and, with one exception that Peter observed, not with several animals feeding at the same time from one carcase, as lions do. Again, this would hardly be practicable with something the size of a mouse, but it would just about be possible if the prey were a rabbit, and rabbits were caught from time to time. Most sharing consisted of one cat feeding until it had satisfied its hunger, then withdrawing and making no objection when another cat took over and finished the meal. Usually a cat would become aggressive to the others while it was feeding, driving them off with a growl or a swipe with a paw, should they come too close.

The cats caught some birds and 'obtained' others, though whether they actually caught them was uncertain. Sparrows were caught fairly often, but the cats tended not to eat them. Whenever a hungry cat did try, the bird was not eaten completely and the cat vomited soon afterwards.

Moles and shrews were caught, but very rarely did a cat actually eat one. Blackbirds, on the other hand, were eaten, and so were pigeons. Perhaps there is something about sparrows that makes them indigestible and unappetising. The cats seemed to think so. Peter recorded one incident, in April:

> The list of prey now includes hedge sparrows (otherwise known as dunnocks, a different species to the house sparrow). One was found lying in the barn, partly eaten and then vomited up again by its captor. Smudge brought in another, played with it and then left it; both Pickle and Domino then played with it, but did not eat it. While one cat was playing none of the others made any attempt to steal the sparrow.

Not all cats are able to catch birds. It is not easy. The cat attack is based on the supposition that the prey will attempt to run away along the ground. Therefore the leap is designed to bring the cat on to the top of an animal, with its front paws extended in front of it. Everything is related to action at ground level. If the prey reacts not by running, but by leaping into the air, the cat that fails to anticipate this will be too low

Smudge with a pigeon carcase; Peter could not tell whether she had caught it herself

and the bird will escape. When hunting a bird, then, the attack must be much higher, to bring the cat down on to an animal that is well clear of the ground, so bringing it back to the ground. It is not every cat that learns to tell a bird from a small mammal, and even if it does appreciate the difference, the solution of one problem creates another. While it is on or close to the ground, the cat is very stable. Its tail balances it, and it has complete control of its body. If it leaps into the air, it has little or no control. The cat throws itself at its prey, and once the attack is launched, the cat can neither shorten nor extend its leap, nor swerve to either side. And if it misses the bird, there will be no second chance – unlike a pounce that fails to strike a running mouse.

The problems of capturing a bird, combined with the comparative inefficiency of a bite to the nape of the neck as a means of despatching it, made the occasional appearance of a wood pigeon in the barn something of a mystery. A pigeon would not be easy prey for the cats, and Peter suspected that the birds had not been caught by them at all, but had been found wounded or already dead.

The failure of the cats to eat shrews is easily explained. The shrew has a gland on the side of its body that secretes a noxious substance, that makes it unpalatable. It takes a desperately hungry cat to overcome the repugnance this causes, but it does happen. Peter noted that on 21 July, 'Pickle arrived in the barn carrying a pygmy shrew, chirping loudly. Smudge walked up to her, apparently attracted by the chirps. Pickle sniffed Smudge's flank and pawed playfully at her face before allowing her to pick up and eat the shrew. Since shrews are not a favourite prey I thought that Pickle had not wanted it, but about half an hour later she turned up with another, played with it and ate it, growling at Smudge and Domino to keep them away.' He added, significantly, 'Food sharing was looking more likely.'

The mole is related to the shrew, and is unpalatable for the same reason. One viewer wrote to suggest that shrews have a venomous bite that might deter a cat. In fact it is only a family of North American shrews that have such a bite – European shrews do not – and the venom is sufficient only to subdue their prey of worms and insects.

Why, then, do cats catch shrews and moles they have no intention of eating? Probably it is because shrews and moles as they scurry through the grass look much like mice or voles. They are killed by mistake, and the fact that they are coloured quite differently from either field or bank voles or fieldmice makes little difference. Colour is not very important to a cat.

In fact it took scientists quite a long time to discover that cats are not colour-blind. The problem is an intriguing one, because the eye of a cat is constructed in such a way that it should be sensitive to colour. The test for recognising colour is very simple. To obtain food the cat must push open one of two or more flaps. These are coloured, and their positions are changed after each feeding, but only one colour leads to food. Many animals very quickly learn the colour that produces results, but when the test was applied to cats they apparently failed to learn. The experiment was almost abandoned as hopeless, with the implication that despite the structure of their eyes cats cannot perceive colour, when suddenly the cats learned the lesson. It seems that cats can see colour perfectly well, but they have never associated it with food. Perhaps this should not surprise us. We eat plants, and we tell edible from inedible ones, and ripe from unripe ones, as much by their colouring as by any other characteristic. A cat eats small animals that move in particular ways. It is the movement that matters, not the colour. So, presumably, no cat has ever identified food by its colour, and apart from those that have been taught to do so in laboratories we may assume no cat does so today.

The acuteness of the cat's eyes is legendary. Indeed, the ancient Egyptian word for cat, *mau*, also meant 'to see'. In the past many people believed you could cure a sty by stroking it with the tail of a black cat, or with a single hair from the tip of the tail of a black cat, drawn across the sty nine times, but the hair had to be plucked from the tail as the new moon was rising in a cloudless sky. Cats' eyes were also sinister, though. One way to become a vampire was to be the first person to meet a cat that had leapt across an unburied corpse – or so people believed in some parts of Europe – and in parts of Britain to meet such a cat would make you blind. The eye was an evil eye, or it could become so. It and its owner were associated with the moon, with the night, and with everything – good as well as ill – that images of the moon and night imply. Cats, after all, can see in almost total darkness, and a light that shines suddenly into the darkness is likely to be reflected by yellow eyes whose owners cannot be seen. The eyes that gleam from gateways and hedgerows into the light of a torch or lantern certainly send a shudder down the spine.

Of course, cats cannot see in total darkness. By definition, 'seeing' is the perception of light, and without light the perception of light is hallucination. However, a cat can see in what we would call total darkness. Its eyes are better than ours in this sense, and the difference arises from their structure. In the first place, the pupils can open very wide indeed – much wider than those of a human eye. They are vertical, and attached at the top and bottom so that they open sideways. Not only can they open wide, but they can narrow to the merest slit, and they can contract and dilate very rapidly. The result is to give the cat's eye what a photographer would call a 'fast' lens, which admits much more light than a human eye can do, and restricts the admission of light more efficiently as well, so that the cat is not dazzled by bright sunlight. To increase the admission of light still more, the lens and cornea are very large. Their total surface – which is the 'collecting area' – is increased by making them much more convex than those of a human eye. Thus the front of the eye collects more light than a human eye when conditions are dim, but as the retina, at the rear of the eye, has not been increased in size, so the light is focussed on to a small area. The result is a bright image under daytime lighting and an image that is clear enough to be of use when little light is available.

That is not all. We can think of light as being composed of photons. When a photon is absorbed by the eye it ceases to exist, because the energy it represents is converted into an impulse that travels to the brain. Behind its retina the eye of a cat has a layer of material consisting mainly of guanine, which is reflective. Any photons that are not absorbed by the retina are reflected back into the eye, where they have another chance to be absorbed. In effect, this layer is a mirror, and it is what makes cats' eyes shine. In darkness, the pupils are dilated widely. A sudden bright light will attract the attention of the cat, but for an instant much of the light that is received by the eye will be reflected back again. Then the cat will either lose interest and look away or its pupils will contract. In either case, the shine will cease and the mysteriously watching eyes will vanish.

Cats' eyes are good, then, but they are not perfect. The buzzard, who was often to be seen sitting on the air as the farm cats set out on their afternoon excursions, could see better, at least in daylight. He had to have good eyes because vision was the only sense by which he could hunt. Birds have a poor sense of smell, but a sense of smell is of little value if you are riding high in the air and your next meal is on the ground. Some birds, and especially owls, have acute hearing, but to use it they must fly low and silently. Hearing is of little value to the buzzard, so he must use his eyes to detect not just small movements on the ground, but the right kind of movements. His vision is very sharp and precise.

The cat can see a great deal, but it sees nothing sharply, especially at the centre of its 'picture'. Things look much clearer at the edges. This is useful, for the cat is able to remain quite still and without turning its head it is aware of all that happens at the edges of its wide field of view. It focuses its eyes much less than we do, but when it looks directly at an object, the image transmitted to its brain is what we would consider blurred.

As with many animals, a direct stare is considered alarming. We regard it as rude to stare at people, and for much the same reason as the cat. We believe we can feel when we are being watched, and the feeling makes us uneasy. It does so because the most obvious purpose of a prolonged stare is to plan an attack. Watch two cats arguing and you will see that they stare directly at one another. Watch two cats greeting one another and you will see that they do not. What is more, a cat stalking prey will usually abandon its pursuit if it notices another cat watching it: while it concentrates on its spring towards the prey it is vulnerable to an attack from the rear, and in most cases it is unwilling to take the risk.

Its blurred vision suits the cat well enough because its other senses more than compensate. Hearing is at least as important to a cat as vision. It hears at least as well as we do, and at high sound frequencies it hears rather better. Even so, many animals have better hearing: the cat cannot compete with the bat, for example. Yet its hearing is good enough to detect the rustle of a mouse that is nearby but out of sight. We could do as well, but the cat has trained itself to note useful sounds and to us the rustle of a mouse is not useful information. Its large outer ears, which it can move, help it to determine the direction and distance of a sound. The farm cats all had large outer ears, as do most domestic cats. In general, mammals which evolved in warm climates are smaller and have larger ears than animals which evolved in cold climates. Body size and the size of the external ears affect the efficiency with which heat is conserved or lost. Because they are well supplied with blood vessels lying close to the surface, ears lose heat quickly. In a warm climate large ears can help to keep you cool, but in a cold climate small ears prevent excessive heat loss. Their large ears suggest that our cats were descended from ancestors who lived in warm regions. The rather smaller ears of the long-haired breeds may be the result of selective breeding, but it also suggests some relationship to a wild cat from colder climes.

The cat also has whiskers, long, somewhat rigid hairs – 'vibrissae' – each of which grows from a root richly supplied with nerve endings. They are sensitive not only to touch, but to small vibrations in the air: a cat can almost hear with its whiskers! Apart from the vibrissae on its face, the cat also has them on the rear side of its fore legs.

Pickle toying with a slow-worm

The cat's face is very sensitive to small changes in temperature, but the rest of its skin is not. The cat whose fur begins to smoulder gently as it lies by the fire feels no pain.

Cats also depend on their sense of smell, though how good this is it is difficult to tell. What we do know is that like many animals, though not humans, cats possess a Jacobson's organ. This is located in the roof of the mouth and is sensitive to chemical stimuli which are taken to it on the tip of the tongue. Whether it is an organ of smell or taste is impossible to say, but the two senses are very similar. It enables the cat to detect and respond to the chemical composition of substances that it samples with the tip of its tongue. This is what the cat is doing when it exhibits Flehmen. This organ is most highly developed in snakes and lizards. These reptiles explore their environment with their tongues, which collect small samples of substances from the air and carry them into their mouths for identification. The cats' Jacobson's organs may be inherited from their reptilian ancestors.

One day at the end of May, Pickle met a slow-worm at the bottom of the field below Peter's hut. She examined it, prodded it with a paw, watched it move, and seemed to be trying to catch it. Peter thought she would kill it, as indeed she could have done easily, but it escaped and vanished into the nettles. A few weeks later, one of the cats did kill a slow-worm, but it was not eaten. No one saw the kill, but Pickle's behaviour was cautious, to say the least. Was she afraid of this legless animal? Cats will kill snakes sometimes, and if they are desperately hungry, eat them. Fear of snakes is widespread among mammals. Once again, we are not alone! Peter wondered whether Pickle's fear was innate. It may have been. The slow-worm is not a snake, of course, and people who fear them are often reassured when they learn that the animal is merely a legless lizard. The distinction is little more than semantic, since the ancestors of snakes also had legs and the large constrictors retain vestiges of hind legs. We are unhappy about animals that have too many legs, such as spiders and octopuses, but even more unhappy about those that have no legs at all. We console ourselves over the inoffensive slow-worm. After all, by calling it a 'legless lizard' we are almost saying that it is an entirely normal animal that suffered some appalling misfortune. Cats do not share our distaste for multi-legged species, but they are wary of snakes, and Pickle was wise to be cautious.

Frogs were rarely eaten either. Domino caught one by the rhododendron patch between the barn and Peter's hut, but she did not eat it and neither did any of the other cats, probably because it was strange and lacked both fur and feathers.

Later on, Pickle brought back a dead weasel. Peter did not actually see her catch it, and if she did the feat was startling. Despite its small size, a weasel is a formidable opponent, and desperation will make it fight savagely, using teeth mounted on jaws that are strong enough to inflict severe injury. The cats became used to catching rats. Perhaps Pickle met the weasel while it, too, was ratting and caught it by surprise. We will never know. What we do know is that she was very proud of it. It had a strong smell, but nevertheless she licked it all over, especially around its tail. Then she left it. Smudge came over and examined it, licking it briefly, then she also left it. None of the cats tried to eat it.

Food was usually brought back to the barn. Even slices of bread and some chocolate cake from Edith's dustbin were carried home and, in most cases, eaten or hidden away for future attention. All cats, including the large ones, prefer to take their prey to the core area of their range – to their 'dens' – to eat it. Such behaviour is very practical, and the cat who dumps its row of dead mice on the kitchen floor is not showing off or seeking our approval. It is simply arranging its larder. While it is feeding, an animal is vulnerable to attack, and cannot be so watchful as it is at other times. While it is carrying food, however, it can remain alert. Therefore the sensible thing for the carnivore to do is to carry its food to a safe place and eat it there.

Hunting left the cats plenty of time for other activities – or inactivities. This was a tribute to their predatory skills. Studies of feral cats on remote islands have shown that an adult cat eats five to eight per cent of its body weight in prey daily. It sounds little enough, but if translated into human terms it means that a 140 lb man would need to eat seven to eleven pounds of food a day. A growing cat eats more – about 9.5 per cent of its body weight a day – and a nursing mother will eat up to twenty per cent. That's a lot of mice! Knowing the cat's appetite, farmers 'employ' cats to control small rodents. But although cats do eat large numbers of small animals (and predators do in general have a regulatory effect on the population size of their prey) the real influence works in the opposite direction. It is the number of prey animals that determines the number of predators an area can support, and not the other way round. Were the farm cats to desert their posts the population of rodents might increase temporarily, but almost certainly it would soon be stabilised again by an increase in the number of other predators, or by emigration as food and nesting sites became more difficult for the rodents to find. Where the farmer scores, is in allowing cats to enter his buildings, where other predators and especially bird predators could not or would not operate, and in choosing a predator that eats mice and rats but not chickens.

Although they ate heartily, the cats killed many animals that remained uneaten. This is something all cats do and the reason for it is something of a mystery. No doubt shrews and moles are killed by mistake and not eaten because they are unpalatable. Perhaps the weasel was killed in a fight, or perhaps it was found dead and Pickle only 'pretended' to have killed it. Why, though, was the frog killed? By no stretch of the imagination could a cat mistake a frog for a mouse. And why are small mammals killed by pet cats which are well fed and often forget all about their victim once they have killed and played with it?

Probably the movements of the animal provide a stimulus the cat finds it difficult to resist, arousing in it an emotional response that must be released. Cats rarely indulge in frenzies of killing as do some hunters including, most notoriously, foxes. In taking one chicken to eat, a fox may well kill an entire flock and leave the rest behind. This behaviour can be explained in rather the same way. The fox enters the run in order to obtain one bird, which it will eat. Its strategy is to choose its target and attack. When it does so, the normal reaction of the other birds is to escape by running. Were they to flee, the fox could not pursue them. Enclosed as they are, however, they cannot escape and so they panic, with a great deal of noise and movement. This reactivates the fox's desire to kill and goes on reactivating it until the noise and movement cease, by which time, of course, there are no chickens left. The birds are the victims of the walls that surround them as much as of the fox.

The cat's 'toying' behaviour with its prey is much more difficult to explain. Other predators do not behave in this way. Some of them may kill more horribly – jackals and dogs bring down their larger victims by attacks to the belly, which is the only vulnerable part of the body they can reach – but the killing is done as quickly and efficiently as the predator can manage. They do not 'torment' their victims. At the end of his year's observation of the farm cats Peter had seen them toy with prey many times, but he was no closer to an explanation of this behaviour than he had been at the start. Some people have suggested that the cat is rehearsing its hunting techniques, and that by toying with its prey it improves its performance. The idea is attractive, but is contradicted by the facts. Toying behaviour is quite different from hunting and killing behaviour, and it is not universal. Some cats play with their prey more than others and some do not play with it at all, yet cats that do not play seem to hunt just as efficiently as cats that do. After seeing the film, one viewer wrote to suggest that the purpose of playing is to terrify the victim, quite literally, so that the victim will vacate its bowels, so making it more wholesome to eat.

Some of the behaviour can be explained. If the cat, for any reason, is unable to kill its victim with a bite to the nape of the neck, it may try to stun it or to break its neck by shaking or throwing it. This could look like toying behaviour, but it has an obvious purpose. If the victim is large, or dangerous, in order to overpower it the cat must first overcome its own fear. To do this it must become highly motivated, and in many cases the killing of the prey does not completely discharge the stimulus it has generated. The cat will work off its emotion by continuing the fight after it has won. It knows its victim is dead, but it may throw it in the air, jump on to it, or even roll on the ground as though wrestling with it. This, too, is behaviour that can be explained, but it is behaviour that affects the dead body of the victim, not the living animal.

No one really knows why some cats, some of the time, allow their victims to escape unharmed, only to catch them again, and again, and again. The only explanation that sounds remotely plausible is that this is infantile behaviour, that in some sense the cat is behaving like a kitten playing with a toy or learning the skills it will use later in its life. Why, though, among predators, is it only the domestic cat that reverts so readily to infantile behaviour?

Could it be that in the course of the domestication of the cat, we have favoured individuals that behave in an appealing, kittenish way, and that these infantile traits also include the toying behaviour that we dislike? Could it be a harmless artefact of the whole selective process that has evolved the domestic cat as we know it today?

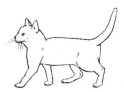

The technology

PETER WAS dripping. Drops fell in a steady stream from the hood of his parka, past his face and on to the front of the coat itself. At its lower edge the parka dripped on to his knees. He cut a dismal figure as he squelched up to the house. His wellington boots were muddy and, hands sunk deep into his pockets, he was as cold as he was wet. His life-style was far removed from that of the white-coated pioneer at the frontier of knowledge who features in so many movies and glossy magazines. His laboratory was the farm, the village, the churchyard and his tiny hut, and it contained no test-tubes or retorts, no computers, no miles of glass tubing arranged in curious, fascinating festoons. Such things did exist, but they were located back at the Department of Zoology in Oxford.

Peter came to the kitchen window and Edith opened it and offered him a cup of coffee. She had become a little wary of inviting him indoors. She could have coped with the mud, but if there was one thing Edith could not tolerate, it was the smell of tomcats. Peter lived in such intimate association with the cats that he was beginning to smell like them. Perhaps one or other, or even all of them, sprayed him as he dozed in the barn? Who knows?

Although Edith and Maurice saw him fairly often, Peter kept to himself for days on end. The cats were most active at night, and so that is when he had to spend time observing them, and he slept during the day. Gradually he became a nocturnal animal himself. Once, late on in the year, he was missing for several days and no one thought anything of it. It was assumed that he was in bed. When at last someone visited the hut, they found he was indeed in bed, but not by choice. He had decided to pick a few apples, but had slipped and fallen from the ladder, twisting his leg and ankle. He managed to crawl back to the hut and into bed, and he lay there for two days before anyone else knew.

His work called for real dedication. Of course, the making of the film provided excitement and Peter enjoyed it thoroughly, but it was an occasional excitement, something outside the daily routine. His first loyalty was to 'his' cats, his job was to observe them almost constantly. Apart from observing what they did, he was also the person most immediately concerned with their welfare, which made him feel responsible for them.

Peter had his relaxations, though. He enjoyed motor bikes and he enjoyed sea angling.

The motor bikes caused much gentle amusement. He had two of them during the year, small machines that his large, gaunt frame – he was well over six feet tall – seemed to overwhelm, making them look even smaller than they were. They buzzed him protestingly back and forth along the road on his cat searches, and he rode them impassively. No flicker of emotion crossed the face beneath the yellow crash helmet. He felt great affection for the machines and valued them highly, so that when he decided to exchange the yellow bike he had brought to Devon, it was with great reluctance that he was compelled to admit that the outside world placed a much lower monetary value on the machine than he did.

On Saturday evenings he sometimes rode his bike to Clovelly, some twelve or thirteen miles away along the narrow Devon lanes. Clovelly, today a quaint tourist haven, was once a tiny fishing port, and it is where Peter did his fishing. The farm cats did not go fishing, so far as anyone knew. Some cats do learn to fish, but the operation is a tricky one. The fish must be caught by a hook of the forepaw and the forepaw must be wetted. Cats dislike water, but less than many people imagine. It has been known for a cat to stand in shallow water up to the tops of its legs and do its fishing that way. The important thing – for a cat – is not to wet the nose. This brings on an attack of sneezing and is uncomfortable. It also ruins the fishing.

Peter might have thought, then, that his trips to the rocky coast would provide a temporary escape from cats. It was not always so. He returned once to describe to Maurice an experience he found almost weird. He was sitting quietly on a rock, minding his own business and that of the fishes, when he became aware of something behind him, watching. Perhaps it was that sensation of eyes piercing into the back, or perhaps he caught a tiny movement from the corner of his eye. At all events, he turned and there, sitting behind him a few yards away, was a row of cats watching his every move. A few enquiries revealed that Peter was not the only angler to use that particular spot. Indeed, it was popular, and no doubt it was the hope of fish that had attracted the cats. Back at the farm, though, there lingered the faintest of suspicions that it was the strong smell of cats, not of fish, that had drawn them to Peter.

It was in November that he returned from one of his fishing trips to find tragedy awaiting him. The project was well advanced and the radio collars had been removed, so there was no way a cat could be traced should it disappear. A passer-by had seen Smudge lying by the side of the lane badly injured, and knowing that the Tibbles household kept an unruly bunch of cats, she had called to tell Edith about the cat. That is how Smudge was found. Had she not been, in a few hours she might well have been dead.

The road was hardly ever used, but Smudge had been hit by a car. The chance that something like this might happen was very remote – it was a factor that was taken into account when the site was chosen – but there it was. Her front right leg and lower jaw were badly damaged and she could have sustained severe internal injuries. As they moved her, very gently, back to the farm, Edith and Peter felt that without skilled help she would die. A decision had to be made, and they had to make it, for both Maurice and Peter Crawford were away filming elsewhere and could not be contacted quickly.

The problem was more complex than it may seem. A strict and very necessary rule of the study was that Peter must never be seen by the cats to interfere in their lives. He was usually there, of course, and he had provided food, but he was always discreet, almost furtive. Except when the collars were fitted and removed, he never attempted to handle the cats. If he did nothing now, Smudge would probably die. If she were taken to a vet for treatment she might well recover, but to help her would necessitate handling her, calming her, caring for her during her convalescence, and so impressing his personality on her in a way that might lead to modifications in her subsequent behaviour. How, for example, could he observe anonymously a cat that had learned to regard him as a friend and that might possibly run towards him in greeting? Would the other cats also be affected by such a change? That, too, might invalidate his work.

The dilemma was theoretical, of course, because Peter was never in any real doubt about what to do. Smudge had to receive treatment, and quickly. Had the colony been larger, he might not have intervened and Smudge might well have died, as she might have done had she been just another farm cat and not a feline film star.

Peter's difficulty over Smudge illustrates one of the dilemmas that he and David faced in the course of the study, and the film-makers had more. How far should a director or cameraman go in order to obtain the pictures the film needs? Some events can be predicted and preparations to film them can be made in advance, but animals are not actors and cannot be directed to perform for the convenience of an audience. Sometimes events must be recreated. Staged events were restricted to those that did not inconvenience the cats themselves. When Smudge recovered, for example, her accident had to be mentioned in the film. Obviously the accident could not be shown, but a sentence in the commentary could be illustrated by a picture of her limping a little as she walked. By the time the film was shot, she was up and about again, and she was limping, so nothing more was required than to persuade her to walk in a particular place.

Edith and Peter took Smudge to the vet who pinned the broken bones and confirmed that in fact she had not been injured internally. The fractures had to be allowed to heal, however, and that meant Smudge could do no more hunting until she was well again. Back at the farm they found a small, disused byre for the invalid. She was installed there, in privacy, and Peter began to feed her a basic diet. When the extent of her injuries was known and treated, and she was installed in her 'ward', Peter telephoned David Macdonald to tell him what had happened and what he had done about it. There had been no acceptable alternative.

Looking back over Peter's notes reveals his special attitude to the cats.

> Until 8.2.78 I pursued a policy of strict neutrality towards the cats. Food was supplied once or twice a day. The cats were observed for a continuous four-hour period each evening and at odd times during the day . . . From 9.2.78 I have restricted access to food so that the cats may feed only when I am present. . . . The relationship between Tom and the small black and white cat (Domino) is, interestingly, much closer than between any other pair. From coat colour, they are probably father and offspring. . . . Warmer weather has meant that the cats sleep close together less often and results for the functional analysis of allogrooming have been less than expected. . . .

From the first day, Peter kept detailed notes of his observations of the cats. By the end of the year there was a thick pile of them, but by that time the method of recording had changed and Peter's life was somewhat easier.

Behavioural studies of this kind are based on direct observation. There is much more to this than simply keeping an eye on the cats and noting obvious events. Births and deaths, copulations, fights and kills

Peter using the Dawkins' organ to record the cats' detailed behaviour

are noted, of course, but if the study is to yield useful information, notes must be made of almost every movement each cat makes. A well-funded large-scale study might employ a team of observers, working shifts, so that the animals can be observed the whole time. The farm study was less ambitious, but it did require Peter to spend hours on end in constant observation. His notes were often detailed:

> On 2.4.78 the three farm females spread out in the orchard leaving Tom in the barn. A little later he walked along the log pile at the edge of the orchard, giving a fairly loud chirping call (similar to the 'seduction' chirp). Smudge and Domino walked over to him and followed him down into the wood where they played, climbed trees and hunted about. Pickle turned up later and greeted Tom with the usual touching noses and flank rubbing.

It is from such notes as these that David Macdonald was able to build up a broad, objective picture of the way the cats lived, but for Peter the writing was tedious. Moreover, he often faced a problem that will be familiar to anyone who has tried to take part in a discussion at a meeting while at the same time keeping the minutes. You cannot write without concentrating on what you are writing, and while you are

The event recorder translated Peter's observations into the language of the computer

writing you are neither observing nor contributing. To help with the routine work, David obtained an 'event recorder', known to biologists as a 'Dawkins organ', after Richard Dawkins, an Oxford zoologist who invented it.

Like all good inventions, the idea is extremely simple. A small electronic organ is wired directly to a tape-recorder so that instead of passing to a speaker system, the signals generated by pressing the keys send impulses straight to the recording head of the tape-recorder, storing the signals on the tape. The tape can then be filled, removed from the recorder, and played back at high speed into a computer which has been programmed to analyse the coded signals. The organ has twenty-two keys, which can be played in combination if necessary, so the instrument is highly flexible. In Peter's case, each cat was given one key for identification. By depressing the key, he indicated to the computer the cat to which the message following referred. Then each of the common events was given a key: nose touching, growling, chirruping, display of aggression, spraying and so forth. He had to familiarise himself with his keyboard, as a touch typist might, but once he was reasonably fluent his observing sessions were spent – in theory at least – in tapping the keys without taking his eyes from the cats. The instrument was powered by batteries and was light enough to be carried around the farm. It made it possible for Peter to produce much more detailed notes more accurately and conveniently.

It was more convenient for David, too, as lengthy words were replaced by the language of the computer.

Why is such detailed work necessary? The first and most obvious reason is that it allows someone else, later, to reconstruct a sequence of events. It is all very well to say, for example, 'Tom caught a mouse', but the bald statement tells us nothing about how Tom caught the mouse, or what the mouse did. Such detail could prove very important, and in ways that might never occur to the observer at the time. You might argue, of course, that every attentive cat owner is familiar with cat behaviour; everyone knows, for example, how cats catch mice. The danger is that only the apparently important events will be recorded at all, and that they will be summarised. This involves interpretation, and introduces a second reason for making detailed notes. How are we to know at the time which events are truly important? There is a real risk that the human observer will accord an order of importance to the events that is based on human experience. Although we share many features in common with them, other animals perceive the world differently, and our priorities are not theirs. It is all too easy for observation to merge imperceptibly with fantasy. So Peter was equipped to record everything he saw, and as much of the cats' activity was at night, David also supplied his field assistant with night vision.

The problem of seeing in the dark was solved years ago, largely on behalf of the army, for soldiers also need to be able to see in the dark. There are two solutions to the problem. The first is to detect animals by small differences between their body temperatures and the temperature of their surroundings. The pit vipers use this technique and have organs – in the pits that give them their name – that are extremely sensitive to temperature differences of a tiny fraction of a degree. Man-made devices, based on the same principle, translate the different heat waves electronically into a picture on a small screen. Thus armed, snipers can shoot accurately at night, but the technique has its daytime uses, too. Electrical engineers use it to examine insulators on poles and pylons high above the ground without having to climb up them. A faulty insulator will be warm, and so it can be detected visually.

The second solution, and the one used by Peter, was simpler still. Although cats can see in very dim light, they are as blind as we are to light beyond the visible spectrum. Torches that radiate only an infra-red light will not disturb them. So David equipped Peter with an infra-red lamp with which he could bathe the cats in a light of which they were unaware. By watching his subjects through special binoculars which were sensitive to these wavelengths, Peter was able to observe their nocturnal activities without disturbing them. The view through these binoculars was monochrome, but it was surprisingly bright and sharp.

The film of the cats included several night-time sequences, with pictures very like those obtained by using infra-red binoculars. Film can be exposed in the infra-red, of course, but this is not how these particular sequences were filmed. Maurice used an 'image-intensifier' rather than infra-red equipment. It gives results that are very similar, but much clearer. Fitted to the front of the camera, the image-intensifier was originally developed for observational astronomers who wished to photograph faint objects. It responds to ordinary visible light, rather than infra-red, but it amplifies the image electronically to produce a picture bright enough for the camera. It cannot be used in total darkness, as infra-red equipment can be, but the night sequences were all shot in moonlight, and this was perfectly adequate illumination.

Peter's other important observational aid was the radio-collar equipment. The technique and technology of this are well established. The functional part of the collar consists of two cylinders lying side by side. One contains batteries, the other a small fixed-frequency radio transmitter whose signal can be distinguished from that of any other transmitter. The receiver is portable and can be carried on a shoulder strap, and has a directional aerial. The tracker is able to select a particular transmission, from a particular animal, and then determine its direction and to some extent its distance.

In preparing the packs, two factors had to be considered. The packs must not be so large or so heavy as to inconvenience the animals wearing them. Experiments with various animals supplied quite detailed information on this point, and there was no difficulty about designing packs that the cats could wear. At the same time, the transmitters had to emit a signal that could be detected over the kind of distances the cats were expected to travel, and their signals had to be emitted constantly for the whole of the tracking period. The batteries must not fade. Packs were put together that satisfied these criteria and they were sent to Devon, but when he saw them Peter Crawford rejected them as being too large. Although the cats could have worn them without discomfort, they would have been very distracting visually, and he insisted that in this instance the requirements of the film be given priority. No other packs were available, and the engineers at Oxford had to make up a set specially. This took a little time, and accounted for the delay in fitting the collars at the start of the project. The compromise transmitters proved adequate and allowed Peter to keep the cats in radio range all the time.

Each pack was painted to match the fur of the cat that was to wear it, and when they were fitted the packs sat snugly against the side of the neck where they could not and did not interfere with any feline activity. Each pack weighed no more than three per cent of the weight of the cat that wore it, which translated into human terms is the equivalent of a load of about four pounds being carried by a ten-stone man. Never once did the cats seem to be inconvenienced by their special collars.

With the radio collars and his receiver, the infra-red binoculars and his event recorder, Peter was kitted out to spy on the cats, but was it only the cats on which he would have to spy? He suggested making an estimate of the local population of those small animals on which the cats were feeding. Without such knowledge, it would be impossible to tell the proportion of the prey population the cats were taking, or to

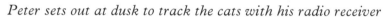

Peter sets out at dusk to track the cats with his radio receiver

calculate the size of the cat population the area might support. Had the study been ecological rather than sociological, it would have needed to continue for several years and the cat colony would have had to breed freely so that a stable balance could be struck between predator and prey. Peter might then have found himself setting traps at strategic points to capture small mammals for counting. Such trapping forms a routing part of much ecological field work.

A technique that is the most common source of information concerning the diet of wild animals is the examination of their faeces. With the farm cats, Peter could see what they were eating most of the time and he seldom had to use this method. Other studies of cat populations have relied on it, and have produced unexpected data about the diets of the cats. In New Zealand, for example, cats were found to be living on rats, rabbits, opossums, mice and stoats, in that order of importance, but at some times of the year they ate very large numbers of insects. These contributed little to their diet, but if a cat met a beetle or a grasshopper it devoured it. Birds, on the other hand, contributed only about twelve per cent of their diet by weight. It may be that the cats had reduced the bird population and that formerly they had relied on them more. On an island off the South African coast, cats fed mainly on a species of petrel that nests in burrows, and on an Australian island their diet included tussock grass and a few other plants. The cat in the wild certainly has a catholic taste – unlike his pampered fireside cousin.

Peter did examine faeces sometimes.

An unburied turd collected on 3.5.78 contained bone fragments from what was probably a mouse of some sort (short tooth row). It could have been the wood mouse I saw Tom eating on 23.4.78, although in that case the maggots must have been living in the dung. Differential flotation in saturated salt solution shows promise as a means of separating the hair in turds from from the interesting bits.

Indeed, Peter's life was very different from that of the movie scientist!

His observations were augmented by his personal notes, and these continued throughout the year. They often included novel ways of looking at the cat. At the end of March, for example, he divided the body of a cat into areas (not literally of course!), allotting to each area an importance based on the way the cats treated one another. 'Records,' he said, 'will show the initiator and recipient of each action, the parts of the body involved, the action, and, except for discrete behaviours, the duration in seconds.' He went on, 'The most important actions are biting, licking, pawing, sniffing and rubbing. . . . If a cat initiates an action by another it appears as the reciprocal. . . . For actions like rubbing, the cat is divided into two large zones.' At that time he did not have his event recorder, and the need for it is very evident: 'Unless the cats get carried away it should be possible to keep up with pencil and paper.'

A few weeks later he recorded:

> The most noticeable thing is that Domino and Tom have interacted much more than usual. Why this should occur is not clear . . . The incidence of neck bites has also increased and on two occasions mounting has occurred. Domino always tries to pull away from bites but continually shows the licking and rubbing which causes them and often carries on with them immediately after a bite has been released. This type of interaction does not occur with either of the other females, although neck bites by Smudge on Domino have been observed during mutual grooming. . . . During some cheek and flank rubbing episodes with Tom, Domino has held her tail erect with the hair frizzed up. This has not been noticed before and the other females do not show pilo-erection, although they do hold their tails up when greeting other cats.

It was the radio collars that enabled Peter to define the ranges of each cat, but Tom was elusive, even with his collar, and sometimes he gave cause for anxiety. Being merely human, Peter had to have a break from work, and early in August he went away for a week. He returned to find that Tom was spending less and less time at the barn. Maurice had seen him on 3 August, he turned up to feed on the morning of the 6th, left in the middle of the morning but came back again in the evening, and visited the barn on the evenings of the 9th and 10th, but did not stay. Almost certainly he was devoting most of his attention to the village, where pilfering was profitable for a cat nimble enough to dodge the occasional boot or bucket of water. By the middle of October, though, his absence was causing Peter concern. 'Tom has not been back to the barn,' Peter wrote on the 14th, 'and despite tracking every night, I have not seen him.' A week later: 'Tom has still not been back to the barn – his last visit was on 4 October. Highlight of a week's tracking was that I actually saw him for about ten minutes last night. At least he looks in good health. As far as I know, he has been ranging to the south of the village but the radio signals have been very erratic and on Tuesday night I could not get a signal anywhere.'

A week later Tom had still not returned, and by this time Peter thought that the cat was becoming truly wild. This would make it extremely difficult for him to approach Tom closely enough to be able to make observations, especially concerning his hunting strategies. Meanwhile, possibly made bold by Tom's long absence, Shadow was seen spraying near the chicken house, and the dark ginger cat with whom Tom had had several fights walked right past the front of the barn, as bold as you please, and sat at the end of the lane. The signs suggested that Tom's absence had become very apparent.

The other cats seemed to know where he was, at least to the extent of meeting up with him away from the barn. With the aid of his infra-red binoculars and directed by the signals coming from their radio collars, Peter was able to eavesdrop on such feline gatherings. One night he spent eight hours watching Tom, Pickle and Domino in a field just doing nothing in particular. Pickle moved away for an hour or so, returning with a mouse, which she ate. The party broke up just as it was getting light, and just after seven o'clock Tom was chased off by a dog.

> He ran down the east edge of the field to the west of the road and then across the road. I found him again at 0712 in the corner of the field. After grooming for twenty minutes he sniffed around, cheek marked, and went to sleep. He was still asleep when I left him at 0800.

So the year wore on and Peter tramped the countryside, sometimes by day but more often by night, his ears full of the bleeps from small radio transmitters, his fingers writing or tapping at keys, often cold, often wet, often bored. Yet his discomfort was endured in the knowledge that slowly a picture was beginning to emerge about the life of his cats. As we sit at home and learn about the behaviour of animals from books or television in the comfort of our armchairs, it is worth remembering the patient hours spent by those enthusiasts who watch and record the everyday activities of their professional quarry. Because of its unique relationship with us, the cat has become the most observed animal in the world, and yet its true nature still remains a mystery.

Social life among the cats

APART FROM Shadow, there were other cats that now and then entered the world of the farm four. Peter, whose head contained an endless supply of suitable names for cats, named them, too, if they turned up at the barn more than once or twice. On the whole, they were not friendly. Shadow wished to be friendly – at least, that is the impression he gave – but not the others. The most aggressive of them was called Minstrel. She would growl and spit at any of the farm cats that came within about six feet of her, and the female farm cats were frightened of her. Tom was not. There was not much in the feline world of which he was frightened. As she snarled and Smudge, Pickle and Domino kept out of harm's way, Tom would simply walk up to her as though he were deaf, or like an old-fashioned policeman who dealt severely with troublesome juveniles. As he approached, Minstrel would continue to threaten, but the threats grew more and more forced, until imperceptibly aggression turned to terror and her legs turned to jelly. She held her ground and on a couple of occasions Tom walked to within about a foot of her. He never touched her. That would have been beneath his dignity, perhaps, but in any case it was unnecessary. The threat was removed, the snarling voice silenced, and Minstrel was put in her place yet again.

Her place was anywhere she chose to be, except among the farm cats. She never caused any real trouble, but at various times, when they were not preoccupied with their own business, the four farm cats would chase her the moment they saw her. She was definitely not liked.

The purpose of the study was to seek clues to the social organisation of the cats, and so of cats in general, and the attitude to Minstrel might provide such a clue. If it was not because of the colour of her fur, or her smell, why did the farm cats drive her away? Perhaps it was she who was unfriendly in the first place, and the farm cats were doing no more than protecting themselves. Yet this would not account for their attitude to Shadow, a much more complicated case. He

72

wanted to be friendly, wanted to join the farm cats and become one of them. The best he achieved was to be tolerated, but he was threatened many times and he was never really accepted. He was like a member of one Scottish clan applying for membership of another. It was as though no mechanism existed for granting his wish. A person can be born into the clan, or marry into it, but a mere change of name will not create the blood ties that link one member to another. Clearly, then, something defined the farm cats as a distinct group and separated it from other groups and other individuals.

The question the study could perhaps not hope to answer, but which it asked, was whether cats are basically social or solitary animals. Do they, like Felix, walk alone, or do they form cohesive groups? What is their true nature?

To see where the mystery lies we must look at a few other animal species. Consider the wolf, for example. As everyone knows, wolves live in packs. The packs are at their largest in winter and in summer they usually break up into smaller packs. A small, summer wolf pack consists of about six males and females and their young. Each year, as new pups are born, most of last year's pups leave to form packs of their own. The pack has a den, a central territory where the young are born and raised, and around this a range that can be very large. The range is defined by the males and females, who scent-mark its boundaries and other key features with urine. So far, the organisation of the wolves is not much different from that which we have seen among the farm cats. Among themselves, the wolves live peaceably enough, but strangers are treated with caution and strangers that wander into the den area are liable to be attacked.

The difference begins with the hunting strategy. Hunting wolves separate, each animal going off by itself in search of game. When it finds an animal, the lone wolf will attack unless the prey is too large for one wolf to handle. Having killed it, the wolf then calls to attract the other members of the pack to share the food. If the prey is too large for one wolf to kill, the wolf that finds it calls to summon help.

Obviously, this is co-operation of quite a high order, and it is typical of canids. Jackals, which are also dogs, live in much the same way. The African hunting dog, too, lives in packs, but it is nomadic and often hunts by night. The coyote has its own version of this behaviour. It is mainly nocturnal, although coyotes can often be seen by day. It mates for life, and family groups sometimes co-operate in hunting by working in relays, individuals taking it in turn to sprint ahead of the group and maintain a lead.

Behaviour of this kind brings obvious advantage. Wolves, jackals and coyotes sometimes hunt animals considerably larger than themselves. At some point in their evolution, some dogs presumably began to collaborate. This increased their hunting efficiency, so the collaborators had more to eat, bred more efficiently and produced offspring that survived, and so in time all their descendants were collaborators because the individualist dogs died out.

According to one current theory, the selective pressure that holds a group of animals together is formed from the genetic material they share. Packs consist of one or several family groups, and this is not coincidence. What matters to the animals is not the success or failure, survival or death of anything so grand and abstract as the species or even the group, but of the individual, provided we are careful with our definition of 'individual'. If we consider each individual as a particular set of genes, then any other individual that shares any of those genes is, to that extent, the same individual. This gives each individual a 'vested' interest in its relatives. Its own survival is still of primary importance, but to the extent that the relationship is close genetically, the survival of a relative is of lesser, but still significant, importance. This requires no great modification of our ideas of evolution by natural selection. In a sense, it is another way of saying the same thing. What it does for us, though, is to remove the difficulty by which individual animals are required to possess some concept of 'the group'. Without one explanation or the other, co-operation is very difficult to understand. We need not expect animals to be versed in theories of genetics in order to apply the concept in their daily lives, for there is no great problem about deciding how animals recognise their relatives and so distinguish between members of the group and strangers. The young know their parents, their litter mates, and members of other litters who are still with the group, and as they mate and become parents themselves, so they know other parents – former litter mates – and the group defines itself by its family relationships. In many species these are reinforced by mutual grooming, in the course of which members scent-mark one another, so the group acquires and maintains a common smell. In cases of doubt, animals that meet can tell in a sniff or two whether or not they are members of the same group. The human equivalent of this might be to require all members of a large extended family to wear an identifying badge. If you look around, you will see that our society is full of badges of all kinds, and among more primitive societies, dress, and body and facial decoration often serve just this purpose.

Having decided how a group may be defined tells us nothing about the kind of group it is. It might amount to no more than an agglomeration of individuals, each of which goes its own way, or it might be a true society, in which members support one another.

Despite its somewhat dog-like appearance, the hyena is related to the cats more closely than to the dogs and it hunts fairly large animals in open country. Can it supply any clues? Of the three species of hyena one, the spotted hyena, is often seen in groups. These groups are socially cohesive, and hunting is often collaborative when the prey is a large game animal. Other hyenas are more solitary, although they will tolerate some food sharing. Among the large cats, only lions live in true groups, and even this holds only for the lions of the African plains. In the days when lions occupied a larger range and habitats different from the parched grasslands with which we associate them today, they lived at most in pairs. The other large cats live entirely alone.

So we are drawn back to the domestic cat, which is traditionally considered a solitary animal with little liking for its fellows It never collaborates, meets others of its kind only when mating, and groups consist only of females and their young. That is the legend, and for many people it seems true. After all, most cat owners own but one cat and that, perforce, lives alone. People who own more than one cat find that the animals usually tolerate one another. Whether they 'like' one another is a meaningless question, since 'like' is a verb we apply to relationships among humans. Not only is this picture the one we have received culturally, it is also what we would expect from studies of cats other than the domestic cat. Why, then, should we doubt it?

An intriguing study of domestic cats, and one that makes us pause, was conducted by Jane Dards in Portsmouth Dockyard, where several hundred cats are contained within high, almost impassable walls. Those cats appeared to be living in groups, the location of the groups being determined by the location of sources of food. From a behavioural point of view, the study was restricted. Jane could not watch the cats in anything like the intimate detail with which Peter and David could watch the Devon farm cats because her cats kept disappearing beneath cover where she could not pursue them. Yet their distribution and behaviour suggested that where abundant food is distributed in discrete locations, cats might form stable, coherent social groups to exploit it.

If you walk round many farms, if you wander through warehouses and factories or watch closely round deserted buildings in the old parts of towns, you will come across groups of cats. They are under no constraint, but they are together. Since the existence of such groups goes some way to contradicting the 'solitary' legend, might it be that more of the legend is suspect? This is what the study of the Portsmouth cats suggests. Are the groups of cats merely aggregations, individuals that have come together more or less by chance, perhaps to share shelter, a source of food, or some other amenity that exists so abundantly that there is plenty for all with no fear of competitive aggression? Or are they, possibly, true societies? Are individuals collaborating? If such collaboration could be detected, it is not only the traditional view of the domestic cat that will need to be revised. Our ideas about all the cats, and perhaps about other predatory mammals, may also need reviewing. What is more, the existence of cat societies, of groups of co-operating, supportive individuals, would accord so well with the 'selfish gene' theory that is advanced to explain the bonds of relationships, that the theory itself would be reinforced. Not only would it be most interesting to discover that cats are social animals, scientifically it would also be extremely neat!

In examining a group of cats, therefore, how might we detect the existence of a society? Clearly, collaboration in hunting would be impossible were the participating animals not bound by a code of behaviour accepted by them all. Each individual must assume a role and, in the example of the wolves, the location of prey requires from the hunter a decision – 'Can I tackle it alone or do I need help?' If animals behave in this way, we may be certain that they live within a social structure. The domestic cats, though, do not live in this way. Even if they did, collaboration in hunting prey that is very small would be pointless. Two cats chasing the same mouse would simply get in one another's way, and were two or more cats to share the eating of a mouse, none of them would get much to eat. The only exception to this which we might expect will occur when a female is teaching her kittens to hunt and eat the catch, or when she is bringing home food to a kitten that is not yet old enough to fend for itself.

We cannot look to hunting and feeding behaviour, therefore. Or can we? The evidence is equivocal. Consider first this brief episode, as recorded by Peter, after the kittens had been born.

On 24 September a rabbit killed by a car on the lane was put out in front of the barn. While a kitten was eating, Smudge joined him and began to feed. Each growled at the other, Smudge left briefly, and when she returned the kitten growled at her and swatted out with a forepaw. Smudge seemed nervous, crouching low and flicking her ears back. The kitten moved away, and when he returned Smudge picked up the carcase and dragged it about two yards away from him. Probably because he was full, the kitten made no attempt to follow, but when I picked him up and dropped him near Smudge, she once again carried the rabbit away.

There is no co-operation or sharing there. On another similar occasion, about a month later, a feeding episode involved several of the cats. Tom was absent at the time.

I gave Smudge a dead rabbit. She ate a little of it, initially growling at the young cat but then leaving him to feed. Pickle stole the carcase from the youngster, Shadow stole it from Pickle, Domino stole it from Shadow, shared it with the youngster and left it for him. Strangely, there was no physical aggression during any of the changes of ownership. Whichever cat was eating the rabbit kept the others away by growling or dragging the carcase about. Presumably the lack of actual fighting is something to do with the likelihood of a third cat making off with the food while the owner and would-be thief battled it out.

In December, one of the kittens born earlier in the year was really beginning to assert himself, but his youth still brought certain privileges, and one of these was the right to share food. It was not a right he reciprocated: he was far less tolerant of the others than they were of him. This tolerance led to one, and only one, incident in which the cats were seen to behave like lions. On 7 December Peter gave Tom a dead rabbit in the barn. Growling to keep the other cats at bay, Tom ate about half of it. The young cat, though, insisted on a share. Tom pawed at him twice, but the youngster kept returning. Tom's growling became less determined, and at last he allowed the young cat to feed from the carcase alongside. For a short time two cats were feeding together, as lions do.

The little drama soon degenerated into farce. Tom finished feeding and seemed to lose interest in the rabbit, so the youngster took it over completely. Smudge tried to take it, but he growled and pawed at her, and he growled at any other cat that came too close. Finally Smudge seemed to present too great a threat to his meal and he dragged the rabbit away from her. This was his undoing. Shadow had sneaked into the barn, unnoticed by the others, whose attention was fixed on the rabbit, and the youngster had dragged the carcase within Shadow's range. Suddenly it was snatched from him, and before he could react Shadow had leapt away on to the bales of hay. Tom had been sitting to one side, grooming himself, but this affront to his authority, and perhaps to his offspring, was too much for him to bear. He launched himself into pursuit, but he was too late. The fugitive vanished, rabbit and all, into a hole, and despite searching Tom could not find him. That was more or less the end of that rabbit. Later Domino found the remains and finished off the parts that were still edible.

Throughout the study all four adult cats would drink milk together from the same bowl. It was only solid food that was defended. It was all very much what one might have expected. Peter felt that what he had seen of the cats thus far suggested that they formed some kind of loosely cohesive group, Tom acting as the leader while he was present.

Peter reported that on 17, 18 and 19 July, food brought to the barn by Pickle was eaten by Smudge. At that time Smudge was 'housebound' because of her damaged leg and could do no hunting on her own behalf. The incident on the 17th concerned two voles. Pickle had been hunting along the road, some way to the north of the barn. She caught a vole and trotted quite quickly with it back to the barn. Peter lost sight of her as she entered the barn, but when he caught up with her she was sitting close to Smudge, and Smudge was eating a vole. It was not certain that Pickle had surrendered the vole voluntarily – Smudge might have stolen it when the two cats were out of sight – and far less that Pickle had caught it specifically for Smudge.

Pickle watched Smudge eat the vole, then off she trotted again, back to the place where she had caught it, and caught another. Again she started back towards the barn, but when she was about halfway she seemed to lose her resolve. She would drop the vole, then pick it up again, and she walked back and forth several times, all the while making chirping sounds and mewing loudly. Finally, she made up her mind, and ate the vole before continuing back to the barn.

Peter suspected that Pickle might have been hungry when she caught the first vole, but that she took it back with her to the barn and then allowed Smudge to eat it. She set off again and caught the second vole, which she also planned to take to the barn. On the way, though, she remembered what had happened to the first vole, and despite her natural urge to take her food home, where she could eat it in safety, her hunger overcame her and she accepted the risk of interruption while eating in the road in preference to the more certain risk of losing her meal entirely when confronted with the needy Smudge. If this interpretation is approximately correct, what Peter saw was food sharing, but it was only partially voluntary.

The next day, however, a possible pattern became clearer. Pickle walked into the barn chirping loudly and carrying a pygmy shrew – not the most favoured of food, but acceptable occasionally. Smudge walked up to her, as though attracted by the chirping. Pickle put down the shrew, sniffed at Smudge's flank and pawed at her face playfully, then allowed her to pick up the shrew and eat it. Peter remained suspicious. Because shrews are seldom eaten, he thought Pickle had not wanted the food, and so raised no objection when Smudge took it. Pickle left the barn, though, and returned half an hour later with a second shrew. This time she played with it and then ate it herself, growling at Smudge and Domino to keep them at a distance.

On the third day, a similar incident occurred. This time the victim was a mouse. Pickle entered the barn with it, chirping loudly. At once Smudge left the hole where she spent much of the time with her kitten. She ran over to Pickle and sniffed at Pickle's head and at the mouse. At last Pickle put it down and Smudge ate half of it. Then, without any protest from either cat, Pickle took over and ate the other half. Domino was present throughout, but took no part in the proceedings. Peter commented: 'It seems very probable that Pickle is bringing in food specially for Smudge despite being hungry herself. The chirps produced by Pickle are considerably louder than those used in other circumstances and Smudge is strongly attracted by them.'

There was one hint more. Some of the fields on the farm grew grass that Maurice sold as hay to a nearby farmer. While the hay was being baled, the three female cats were seen in the field. They sat a little apart from each other, forming a triangle at the edge of the field, and they were watching carefully for such small animals as might come their way. This could not be called collaborative hunting, at least not by intent, but it was clear that because all three cats were in the same area, the chance of prey being caught by one or other of them was increased. A small animal might evade one cat, but only by moving within the range of another cat. Was the arrangement deliberate? Probably not, but it was effective all the same.

Such food sharing was the first positive evidence the study produced to suggest that under certain circumstances cats are capable of co-operation and altruism. There was what might have been a hint early in the study, when Tom and Smudge showed interest in Pickle's distress over the fitting of the dummy collar. Tom and Smudge might have been expected to run away from possible danger, for Pickle's screams must have suggested danger. They did not. They approached as though to investigate and, perhaps, even to help. 'Not what would be expected from a solitary carnivore,' Peter noted.

We mentioned a little earlier Jane Dards's study of the ecology and behaviour of a large colony of cats in Portsmouth Dockyard. These cats are interesting because although they are no strangers to human company – some ten thousand people work in the dockyard, and many of them feed the cats and provide them with boxes and bedding for shelter – they are isolated from other cats and have been so for a long time. On two sides the dockyard is bounded by the sea, and on the remaining two sides by a high wall. There was a walled dockyard on the site in the thirteenth century, and the present walls were built in 1711 and 1864. It is possible, therefore, that the cat colony dates back to medieval times, but more likely that it dates back one or two centuries. At all events, it has been there for a very long time, and since the animals that founded the colony became trapped inside the dockyard area – they may well have come ashore from ships – it has been virtually impossible for cats to leave or for new cats to enter. This was mentioned in the film, and some viewers found it extraordinary that there is such a thing as a wall a cat cannot pass. This is not true. Cats have difficulty in descending from great heights, so it is clear that if a wall is high enough to make ascending it difficult, it will become impassable. A cat that did manage to scramble to the top would almost certainly be marooned. The gates are a different proposition, but only to a limited degree. To pass through a gate a cat would have to cross a large expanse of open territory, which would make any half-wild animal nervous, and except during working hours, when humans and their vehicles are passing through the gates constantly, the gates are kept closed and form a barrier as formidable as the walls. It would seem likely, then, that even if a few cats wanted and managed to pass into or out of the dockyard, in the course of two centuries or so very few of them have done so. To all intents and purposes the colony is cut off from the outside world.

The dockyard supplied the cats with plenty of food and shelter, so the environment supported a density of about two adults per hectare (rather less than one an acre). In a rural environment you might find two cats ranging over about 16 hectares (40 acres). The hospitable environment and consequent high population density meant that home ranges were generally small and ranges, even of toms, overlapped extensively. The females tended to share family-group ranges. This implies a social form of behaviour, since if ranges are to be shared by a group, a degree of co-operation is required to prevent conflict. Jane suggested that although the cat is usually solitary, such range sharing was a response to a very favourable environment. However, she went on to make an intriguing comment, pointing out that some scientists have suggested that among mongooses – animals very distantly related to the cats, being members of the same sub-order but not of the same family – a shortage of vertebrate food leads to feeding on invertebrates so that the animals can continue to live in groups. Thus, if a social animal can modify its feeding behaviour in order to maintain its social way of life in a poor habitat, might it not be possible for a solitary animal to adapt its social behaviour and become sociable in a rich habitat? Her point was that by adapting to a social way of life the cats were able to exploit their environment more efficiently. This is relevant to the study of the farm cats in so far as it provides at least one reason to explain sociable behaviour in otherwise solitary animals. Compared with the dockyard cats, the farm cats were living in an inhospitable environment.

There have been studies of groups of cats that aimed to discover whether a social hierarchy emerged, or could be made to emerge. One such study, done in America, created conditions for cats that were so unnatural that the results must be suspect. Three groups, each of four cats, were trained to obtain food by pawing it from a dish placed outside their cage. Then the cats were made to compete for a single piece of meat provided for each group. In each of the three groups a dominant cat emerged, but that was the end of any hierarchy. The remaining cats were equals, and the dominant achieved and maintained its position only because it was better at obtaining food than its rivals. The dominants did not seek to extend their dominion into other fields of activity and appeared reluctant to fight contenders. In a somewhat similar, but less contrived study, cats that were made to live together as a group produced two dominants, who took the most highly preferred food and the most comfortable sleeping places, but their dominance was not asserted strongly and the group never became a tyranny. Should a subordinate cat curl up to sleep in the place usually reserved for one of the dominants, it was not disturbed. When the cats were hungry and food was supplied, they all ran to the food and ate together, but then the subordinates stood back and allowed the dominants to continue and so to eat more.

Tom, although at times he was clearly in control in the barn, was never a tyrant. At times he intervened to defend one of the females from a stranger, and the females were often less nervous when he was present, but displays of aggression towards the females were never attempts by Tom to assert himself. They were sometimes rebuffs

Tom clearly enjoys Domino's attentive grooming – but he was not always so responsive to such approaches by the females

directed towards cats that might have robbed him of food while he was eating – and all the cats behaved this way and Tom was not immune to such displays of aggression from others. Sometimes they were displays of irritability. Tom's temper was uncertain at times. Again, there was no reason to suppose that such minor outbursts had anything to do with Tom's social position. He was in a bad mood and wanted to be left alone, and that is all there was to it.

Yet there were indications that the farm cats formed some kind of group. Consistently they showed aggression towards intruding cats, and they did so even though the intruders must have been well known to them and, at least in some cases, were not being aggressive themselves. The farm cats seemed to regard themselves as a distinct group whose members were accorded rights and privileges not accorded to non-members. Mutual grooming was a sure sign of membership.

Perhaps such 'group solidarity' as the cats displayed can be explained by their close familial relationships. Behaviour that promotes the welfare of part of the group at the expense of the individual needs a different explanation. An altruistic act is one that confers benefits on the individual towards whom it is directed, but none on the perpetrator of the act, who may actually run a risk or make a sacrifice in order to perform it. Many animal lovers believe such behaviour can be explained simply, but simple explanations are based, almost always, on the supposition that non-humans are really humans that just look different. The supposition is dangerous for several reasons. In the first place, it cannot be tested. In the second place, non-humans live very differently from humans, and if we believe that successful species are more or less well adapted to the circumstances under which they live, we must allow that these differences will be reflected in behaviour. Finally, the anthropomorphic supposition implies that humans are superior.

But there is clearly an 'animal nature' that we share with most mammals and that requires certain responses in certain situations. Attacks on our spouses and children, for example, are likely to cause us to intervene or at least to want to intervene. Our laws exist, in part, to prevent people from 'taking the law into their own hands' in such situations. No matter how we may rationalise what we feel – and the feeling is very strong – our children are the next generation of ourselves and our spouses are the means whereby we ensure that such a generation can exist. Among species that mate for life, and among all species where the defence of the young is concerned, such intervention is to be expected. Why, though, should we court injury by going to the defence of a total stranger? As humans we can explain this, but in order to do so we must refer to the existence of human societies and the advantages their members derive from the maintenance of stability within them, which requires that aggression be opposed. When we see such behaviour in a non-human species, must we also suppose the existence of a society and a comprehension that internal violence is socially undesirable? Probably we must, but the supposition is a very large one and so far as we know for most species it is uncertain, since we can never be entirely certain that the victim to whose aid the altruist goes is in fact a stranger. It is more likely that the threatened member of the group is also a blood relative. There are exceptions, possibly, among some of the primates, but in general it is safe to assume that animals bound by social bonds and an apparently organised way of life are also members of the same family.

Animals can be deceived, so that as a result of human manipulation the bond is formed among individuals that are not closely related but that would be in a wild setting. The most obvious example of this is the flock of domestic chickens. They will establish a 'pecking order' hierarchy and so a society of a limited kind, even though they do not comprise a single family. Co-operation among them may extend to the defence of core territory, and it is not unknown for a bird to use its body to keep warm another bird that is virtually naked because it is in full moult. This is not truly altruistic behaviour, however, because although the protective bird derives no benefit from its behaviour, neither does it suffer any real inconvenience.

We can see, then, that the question of the existence or non-existence of a society among the cats was at this stage a very tantalising question. Peter observed that the cats lived together amicably, that when he was present Tom played a protective role that was recognised by the females, and that members of the group seemed to recognise a difference between their fellow members and cats that were not members, even though those cats were well known to them.

As the study developed, attention turned towards Shadow, who seemed to provide the key to an understanding of the social structure of the group of farm cats. For the scientists, he was something of a bonus, a cat whose participation had been neither planned nor predicted, and who arrived apparently of his own volition. It seemed that he wanted to join the group, but his overtures were rebuffed. This is what the scientists would have expected. If the group was held together by familial bonds, this would determine the attitude of its members towards all intruders. It was no less difficult to account for Shadow's persistent attempts to win acceptance. The farm cats had food and shelter, which he wished to share.

Little by little, though, he came to be tolerated. It was in March that he and Tom found themselves drinking milk from the same dish, to their mutual horror. In April Shadow was hunting, although he continued to visit the barn. Peter discovered the hole among the hay in the barn where he slept. The aggression between Shadow and Tom continued, although Tom never sought to press home his attacks when strategically he could have done so, despite the fact that Shadow was inside the barn itself. By the end of April Peter noted that 'Shadow appears well established in the barn', but Tom still sought to assert himself now and then. 'Tom approached, Shadow backed away, crouched and hissed, showing the fear-threat face. Tom turned, spray-marked the entrance to Shadow's hole and walked away. Usually Tom ignores Shadow.'

Tom and Smudge resented his presence, but on 5 May Peter saw Domino lick his neck when he joined her to drink from the dish of milk. Tom was acting oddly, forcing fights with Shadow by cutting off his retreat when the intruder would have avoided a confrontation, and by walking into his attacking position rather than rushing and so gaining the advantage of surprise.

This inconsistency continued for some time, so that Shadow never knew whether or not he would be attacked. The other cats left him in less doubt and Pickle was particularly hostile towards him, although even their behaviour was not really consistent. Sometimes they would attack, sometimes they ignored him. Shadow was frightened of Tom and nervous of the others. He never retaliated when he was threatened, always preferring to retreat and escape. At the very end of May, though, Tom went away for several days on one of his excursions. Shadow seemed less frightened and with Tom away the females were more tolerant of him, although one day Pickle chased him back into his hole three times, as soon as he showed his face.

Then, at the end of June, Peter saw an incident that puzzled him.

Shadow decided to deposit a turd in the passage at the west end of the barn and was disturbed in the middle of proceedings by the approach of the three farm females. Smudge and Domino stood about three yards away while Pickle walked up and almost touched noses with Shadow who stood, or squatted, his ground. Even with three potentially hostile cats within three yards of him, Shadow buried the turd before running off as usual. It appears that Shadow is considerably less frightened of the farm females than all the chasing about suggests. Group aggression is the only explanation which readily springs to mind – if Shadow ever got caught he would have to fight, possibly against more than one cat, so he runs away whenever possible. If fighting ability depends mainly on size, he would be able to defeat any one of the farm females.

By autumn the farm cats were quite tolerant of Shadow, who came and went much as he pleased. He was still not accepted, however. One day, at feeding time, Tom took no notice of Shadow, who sat for some time about ten yards from him, then walked across the front of the barn, while Tom watched placidly. Then the kitten appeared and rushed up to Shadow. Shadow stood his ground, but it was too much for Tom, who got up, stalked Shadow, and then chased him up to the top of the bales, where Shadow escaped into a hole. Even then Tom did not abandon the attack, making three fierce attempts, tearing with his forepaws, to get into the hole. Finally, he scent-marked above the hole before walking away. The most reasonable explanation for the attack was that Tom reacted to Shadow's failure to run away from the kitten as a threat to a member of the group from a non-member.

Even this incident did not deter Shadow, and by mid-September his relationship with the others was almost friendly. Aggression towards him ceased almost entirely, and once Pickle rubbed his flank. Shadow sniffed at the tail end of both Pickle and Domino on different occasions, and this friendly gesture was not rebuffed. Once he even went so far as to lick the kitten's head, briefly.

Smudge took to bringing in food for the kitten, but sometimes Domino arrived on the scene first and Smudge gave it to her instead. On one of these occasions Shadow was first to arrive. He sniffed at the vole she had brought, but Smudge held it away from him and when Domino arrived dropped it in front of her.

That happened late in September, but by the third week in October, Peter was recording that Shadow could start a game with the kitten without fear of interruption from the adults. Clearly his status was changing. It was not long after this that the ginger cat from a nearby farm invaded the barn. The kitten and Domino acted aggressively, though cautiously, since neither of them could have hoped to defeat him, while Shadow sat nearby, the hairs on his tail erect. It was as though by this time he considered himself a member of the farm group, and shortly afterwards he scent-marked near the chicken house.

By November Shadow was looking actively for females in oestrus. This produced a few interactions, and probably accounted for his increased nervousness when Tom was present, but otherwise his position among the farm cats seemed secure and he was treated as a friend. It looked as though his persistence had brought him success in the end.

The development of this relationship was puzzling, for if it was possible for a cat to join a family group from outside, the theory that it was familial bonds alone that united the group was weakened. For much of the time Shadow was a juvenile, and his immaturity meant that Tom in particular may not have regarded him as a sexual rival, so he was treated less roughly in the early stages than he might have been. This partial tolerance combined with Shadow's persistence led to a situation in which Shadow had been on the edge of the group for so long that it seemed to the cats as though he had always belonged to it.

In the end we cannot be sure. Shadow was accepted socially, but whether this amounted to full acceptance as a group member was not clear. All we can say is that the old image of the solitary cat was severely damaged. It cannot be proved yet, but the study had already provided evidence to suggest that under certain circumstances, created most probably by the way in which food is distributed about the immediate environment, cats may form stable groupings in which members distinguish between fellow group members and non-members, adopting a different attitude towards other cats depending on whether they belong to the group or not. Probably the group is united by familial bonds – genetically – but the history of Shadow the 'outsider' suggests that it may be possible for these bonds to be extended to create a category of 'associate member'. Shadow, through his persistence, became tolerated if not entirely accepted by the family group.

What is a cat?

PICKLE HAD been sitting beside the hedge for what seemed to Peter a very long time. The place was familiar to both of them, but their view of it was different. Peter saw it as a location in a landscape, a place much like many others, and his knowledge of it was mainly visual rather than aural. It was the spot where the summer bellringers' practice echoing from the church tower arrived with a particular quality subtly different from its quality elsewhere. It was the place where the wind sighed through an old elm tree with a characteristic murmur and where the sound of an occasional tractor driving up the road, though muffled by the vegetation, was still clearly audible. Above all, it was the place from which you could see the back of some old farm buildings, from which Maurice's farm and the cats' barn were hidden by the trees, from which the field appeared in a particular shape, created by the perspective from which it was viewed.

To Pickle, too, it was a place in a landscape, but her landscape was different. Visually it was much smaller, much less clearly delineated, and lacking colour, of which she probably took no notice. For her there were no broad vistas – at least, no vistas that Peter would have called broad. She lived so close to the ground that any herb growing more than a foot tall would obstruct her view. Her horizons were close, but far from presenting her with a problem, the fact was to her advantage. The plant that obstructed her view also concealed her.

She knew her world as a place of textures, smells and sounds, and it became important visually only when sudden movements alerted her to the presence of food or danger. Her awareness of her surroundings was intense and intimate and it reached her almost through the pores of her skin. The pads on her feet felt the ground on which she walked, distinguishing clearly between one surface and another, and the whiskers on the backs of her front legs elaborated on this information to tell her about the depth and density of the vegetation through which she moved. The sounds she heard helped her to orientate herself, for to the direction of those made by humans, and those made in the trees far above her head, she could add the gentler rustlings of the wind through the grass and along the bottom of the hedge, which were more reliable because they were more constant.

Although dominated by her large eyes, her face was a complex mass of sense receptors and vision was not her most important sense. Her delicate, moist, pink nose could detect small changes in the air. This enabled it to monitor the wind direction whenever it was not required for more pressing business. It could locate small areas that afforded shelter from the weather and which might prove attractive to a small animal for that reason. It could detect the presence of an object that was just a little warmer than its surroundings. Her whiskers told her much more than whether or not her body would pass through a particular opening. They picked up and amplified vibrations of all kinds, including sounds, and when she needed to bring her mouth to the place where it could seize food, her whiskers could help to guide it with the precision of the controller who guides an aircraft along a predetermined glide path, but with none of the fuss. It was not a skill about which she had to give a moment's thought.

It is easy to think of a cat as an animal that is part human. Indeed, it is so easy to do so that throughout the centuries the cat has been regarded as a benign or malevolent, but intelligent and essentially human, spirit. So it has been revered and persecuted. It is easy to think of the cat as human because it shares certain characteristics with us. Its large eyes resemble human eyes, and so we imagine they express emotion, as human eyes may do. The eyes form part of a face that is different from a human face, but that is a face nevertheless, with ears, nose and mouth all in their proper places. Yet the similarities are superficial, the human attributes no more than our fancy or our vanity. The cat is not a human.

This becomes even more apparent when we consider the way the cat obtains information about the world in which it lives. Its eyes are large, as ours are large, for what we may think of as an engineering reason. If they are to work efficiently, they must contain a certain number of components. The size of the components is determined by the size of individual cells, and so to some extent the size of the eye is limited. Were it much smaller than it is, it would not perform the tasks allotted to it. Since it accepts light signals it must use a lens and, as with any lens, the distance between the lens and the retina, on which the light is focused, cannot be varied beyond rather narrow limits. The eyeball must be more or less spherical, for how else can it be given sufficient mobility to be of use? It is not the cat, though, but the owl that demonstrates such engineering constraints most dramatically. It, too, needs efficient eyes, and so its eyes, too, must be of a certain minimum size. The owl, though, has a small head, and so the eyes occupy a large part of the available space. Even so, there is not really room for the eyes in the head of an owl and something has to be sacrificed. What is lost is the musculature that would allow the animal a wide range of eye movement. There is simply not room enough for both eyes and all the muscles and points of muscle attachment needed to move them. The result is that although the owl has very large eyes, it has very little ability to move them. It stares fixedly, and when it wishes to look in another direction it turns its head. The adaptation is effective, for by shifting the necessary muscular emphasis to the neck the owl is able to turn its head almost through a half circle. It can gaze fixedly to its own rear!

The possession of eyes that resemble human eyes closely does not imply a similar use of them. We are descended from ancestors who lived in forests and many of our surviving, though distant, relatives are arboreal to this day. An animal that lives in trees has a different need for eyes. It must judge distance very accurately. So must a predator, but its judgement need not be quite so precise, for it can make fine adjustments during an attack in a way that is not possible for an animal that is leaping from one tree to another. Despite their prodigious gymnastics, falls from trees are not uncommon causes of injuries and deaths among gibbons. The evolutionary advantage, then, must go to the animals that have good stereoscopic vision, not because this will help them find food, but because it will help them to move safely about their habitat. The tree-dweller is much more dependent on eyesight than the ground-dweller, and the reason that we obtain most of our information visually is that our remote ancestors lived in trees. We have inherited not only their eyes, but their ways of using eyes. The cat and the owl have evolved by different routes. Their eyes are similar to our eyes in construction, and have a similar evolutionary history, but for most of that history they have been used differently. We see colour, and use the information we derive from our perception of colour, because in the forests perception of colour was useful and because, like our ancestors, we are diurnal in our habits. The owl has no use for colour, since it is mainly nocturnal. The cat perceives colour perfectly well, but ignores it because it is of no importance. Vision itself is less important to cats than it is to us.

From his vantage point some distance away, it was fairly obvious to Peter why Pickle was sitting by the hedge. Earlier that morning he had seen her catch a vole close to the spot where she was sitting now. She was waiting for another.

To Pickle the waiting was not on the off-chance that a vole might happen to pass. Luck played its part in her life as it plays in all life, but her hunting strategy was not based on hopeful ignorance. She knew precisely what she was doing. It was true that she had caught a vole by this part of the hedge a couple of hours ago, but that fact alone would not justify her long wait. It would bring her back to the spot for a further investigation, but the purpose of the investigation would be to determine whether or not other voles were still present. She would not wait for a meal that did not exist. As it was, she had examined the hedge critically. She had seen the small burrow entrance, of course – even a human could have seen that – but she had done more. She had detected the faintest, most tenuous sounds, squeakings and rustlings that she could tell apart from those made by plants in the wind, sounds that told her that the voles were still at home. The place smelled of vole, too, and the smell was fresh. One or more voles had passed by within the last half hour. So she sat, still and silent. She gazed into space, poised and yet perfectly relaxed. She did not stare at the hole, as a human might have done, but gazed, her eyes unfocussed, in a direction about seventy degrees to one side of the hole. Had she looked at the hole directly, her inability to focus sharply on objects at the centre of her field of vision would have made it possible for an animal to pass too quickly for her to react. As it was, she could and would react instantly to the slightest flicker of movement at the corner of her eye, at the mouth of the hole. She sat waiting, for she was fairly sure there were voles for the taking and that sooner or later one of them would venture forth.

Did she think as she sat there? Was her mind as unfocussed as her eyes? It was enough for her to feel the environment around her and to be part of it. She could and did experience emotions, but she did not think, as Peter was thinking. Perhaps she had memories. Much of the skill she practised every day of her life had been learned, and she remembered that, so to some extent she had memories. If she was remembering, as she sat by the hedge, her reminiscences were not of the human kind.

Peter, on the other hand, was thinking. His eye had been caught by Pickle's radio collar. It was working correctly, for it was the signal from it that had directed him to where she was now. As he looked at it, though, he remembered the rehearsal with the dummy collars that had sent Pickle into a screaming frenzy. She could be handled only with great difficulty. Other cats that live apart from humans cannot be handled by humans at all. In that sense they are completely wild. Why is it that some cats are tame and have no fear of humans – even approaching strange humans to be stroked, on the open, neutral ground of a city street – while others will hide at the approach of a human?

If you go to the highlands of Scotland and are very patient or very lucky, or both, you may catch a fleeting glimpse of an animal that looks rather like a domestic cat. It is on the large size, and tabby-marked, but its tail has conspicuous ring markings and is thicker than that of a domestic cat. It might be a feral domestic cat, from its appearance. In fact it is *Felis sylvestris*, the truly wild cat of northern Europe and Asia. It is a close relative of the domestic cat, *Felis catus*, and can interbreed with it. Some people have suggested the two have interbred to such an extent that all the wild cats alive today contain some domestic cat in their ancestry. It is suggested, too, that the retreat of the wild cat, whose range is now confined to the most remote and inhospitable regions, is due to its inability to compete against the more successful domestic cat.

From time to time people have captured wild cats and tried to tame them, without success. The animal simply will not tolerate humans or human dwellings. It spits defensively if approached, fights anyone who attempts to handle it, and if it is confined to a house it cannot be trained to use a toilet box or be restrained from causing mayhem. Even young kittens are untameable. Like all young mammals they can be handled safely and so appear tame enough, but as they mature their wild nature asserts itself and they become as intractable as their parents. They forget the human ways they have been taught, and in the end they must be liberated to survive as best they can – which may not be well if they lack proper feline education – on the craggy mountain sides, or be donated to a zoo where they will pace out their lives neurotically behind bars through which stare the humans they fear so deeply.

The two species are so close to one another, so similar, and yet so different in their behaviour towards us, that for many years people wondered how one had become the other. How was it that our forefathers managed to tame *F. sylvestris*, while we have failed so signally to do so? The answer, of course, is that despite their similarities the domestic cat is not *F. sylvestris* and never was. It has come to us by a different route. Who then is this familiar stranger?

Cats are carnivores. This is more than a simple statement of observable fact; it has a meaning in terms of the classification of species, and that classification may help us. The order Carnivora consists of two sub-orders, Aeluroidea and Arctoidea. The former embraces the cats, hyenas and civets, the latter the dogs, weasels, bears and raccoons. Within its sub-order, all cats belong to the family Felidae. This suggests that while they share certain features in common with, and may well have descended from the same ancestors as the hyenas and civets, it is possible to distinguish a cat from its non-feline relatives. All cats, large or small, are constructed in much the same way, so that it is possible to examine an animal anatomically and to say at once whether or not it is a cat. It is possible to tell, too, to which of the four possible genera it belongs. The cheetah differs from all other cats in several ways, the most striking of which is its lack of retractile claws, and it is placed in a genus of its own, *Acinonyx*. The next important division is between the large cats and the small cats. Unfortunately, though, the division cannot be made by size alone. All large cats are large, but some 'small' cats are large, too. In 1916, R. I. Pocock, of the Zoological Society of London, observed differences in the structure of the hyoid apparatus between groups of cats. The hyoid apparatus is a series of connected bones situated at the base of the tongue. Animals with a fully developed hyoid are able to purr, but they cannot roar. Most of them also happen to have vertical pupils to their eyes, although there are exceptions. Cats whose hyoid apparatus fails to develop fully cannot purr, but they do roar, and all of them also happen to have round pupils. These – the 'roarers' – Pocock placed in the genus *Panthera*, and they include the lion (*P.leo*), the tiger (*P.tigris*), the leopard (*P.pardis*) and the jaguar (*P.onca*). The 'purrers' he divided into two genera, *Felis* and *Lynx*. *Lynx* includes, obviously, the Old World lynx (*L.lynx*), the Canadian lynx (*L.canadensis*), and the bobcat (*L.rufus*). The other 'purrers' belong to the genus *Felis*, and apart from the Old World wild cat and the domestic cat the genus includes *F.concolor*, the puma or mountain lion, which is the size of a leopard, and the ocelot (*F.pardalis*), which looks like a leopard. There are many members of *Felis*, though, that really are small in size, and it is not always easy to tell one from another. In all, there are thirty-four members of the Felidae family, and most of them are small.

Where you find a family of animals that is very distinct from all its relatives, and highly specialised for a particular way of life, it should come as no surprise to discover it is of ancient lineage. Dinictis, for example, was a cat very like a modern cat, although its brain was rather small. The animal itself was about the size of a lynx, it had retractile claws, cat-like teeth, and it lived in the Oligocene epoch, which ended twenty-six million years ago. In fact, animals that undoubtedly were cats have been traced back forty million years, to a time when other mammal families were barely distinguishable from one another. The cats, then, are among the oldest of mammals. They appeared early, were successful, and have prospered ever since. Modern cats appeared about ten million years ago, since when they have changed very little.

It was a very long time, though, before the cat became domesticated. It seems most likely that it was the Egyptians who first kept cats, and the cats they kept were adapted not from *F.sylvestris*, but from the local African species, *F.libyca*, an animal that is rather more lightly built than *sylvestris*, though still larger than the domestic cat – about three feet from nose to tip of tail – and that has a coat usually of a solid colour ranging from light brown to sandy. The coat colour varies from one part of Africa to another, and in some places *F.libyca* is grey, faintly striped, with a white chin and bib, and its close relationship to the domestic cat is very evident. It is said that the modern Abyssinian breed is the cat that resembles most closely the cats that were kept by the Egyptians, and this may well be true, but *libyca* could pass unnoticed in any European city street, were it not for its size. It has been suggested that the tabby marking which is so common in domestic cats, especially those in Europe, is the result of hybridisation between imported domestic cats, descended from *libyca*, and indigenous *sylvestris*. This may be so, but animals known to have been born from such matings are invariably too wild to make satisfactory pets.

F.libyca can be tamed fairly easily. It is much more tolerant of humans than is *sylvestris*. However, a tame *libyca* is still far removed from the docile pet we know today. One viewer wrote to describe the behaviour of a cat she had owned in Africa that had been half-domestic, half *libyca*. It was fairly tame while at home, but outside the house its behaviour was very different. 'He swiped food from his plate, ran off and "killed it", fought neighbours for steaks they were preparing, and when we left him for six months he took to the bush. . . . (he) would come shopping with me and attack all dogs . . . He delighted in tormenting our cow . . .'

Clearly, then, something happened to the tamed *libyca* to change it into the domestic cat we know today. But what? No one really knows. It is one of the many questions about cats to which we have no answers. We can speculate, though.

Pickle, sitting beside her vole burrow, is clearly a different animal from either *libyca* or *sylvestris*. She looks different, is smaller, and her behaviour is quite different. Despite the difficulty Peter had in trying to handle her, she could be tamed, and kittens she bore could be tamed very easily and would remain tame. *Felis sylvestris* has only one litter a year, but Pickle could have several. She is sufficiently different to be classified as a member of a distinct species, *catus*. The change, then, was genetic.

It is usual for animals to undergo important changes following their domestication. Some change more than others, but our cattle and sheep, for example, are quite different from their wild ancestors, and it is not merely the result of selective breeding for high rates of production. The aurochs, from which our cattle are descended, was large, secretive, and very dangerous. Indeed, it was such a difficult animal to deal with, that for many years historians wondered how it had come to be domesticated in the first place. Why would people bother with an animal that was plentiful in the wild, and that must have been a disaster in the village compound? The answer was found, or at least an answer that is plausible was found, and it has a relevance to Pickle and her friends. What appealed to early farmers was not the animal as a source of milk, meat or hide, but its horns. On the aurochs these were long and curved outward to either side of the head. They were wicked weapons, but they were also reminiscent of the crescent moon, and the moon was a potent religious symbol. The cattle, then, were of ritual importance – as they were much more recently in India – and it was for this reason that they were held captive. After many generations of domestication they became smaller, as cats have done. The reason for this is not at all clear. It has been suggested, for cattle, that in the early days, when the stock was built by capturing wild specimens, it would have been small individuals that were favoured because they were less dangerous to handle, so that from the start the domestic variety was selected for its small size. This may have been so, but it seems unlikely. A small individual is no less dangerous than a large one – and its size says nothing about its temperament – and it is more likely that the most frequent captures would have been of calves, whose mothers had been killed. In any case, cattle are not the only species whose size has diminished following domestication. With sheep it was principally the appearance that changed, as the appearance of cats changed. Among sheep, the effect was more dramatic, for the coat of the animal altered drastically. Wild sheep, like many animals, have a coat consisting of two layers. Next to the skin the hair is short, curly, and useful for trapping a layer of warm air. Outside that there is a second layer of much longer guard hairs that protect the inner coat and help to keep it dry. No one has the slightest idea how it was that little by little domesticated sheep lost almost all of their outer coat and expanded their inner coat to compensate, so creating the woolly fleece, but that is what happened. They were brown sheep, of course. Our white sheep are quite recent, and the result of selective breeding to produce a wool that is easy to dye.

Genetic changes, then, are quite usual as a result of domestication, and since such changes are likely to occur as a 'package' affecting several characteristics together, we should not be surprised to find that those aspects of behaviour that are inherited are among the characteristics that change. While we have some knowledge of the mechanisms involved in genetic changes that manifest themselves physically, we know less about genetically determined behaviour. It has been observed, though, that the behaviour of an adult domestic cat living with humans resembles in several ways the behaviour of a young kitten. It enjoys being handled, and a kitten finds comfort in the close proximity of its mother or other adults and often fails to distinguish between them and mammals of other species. It is playful, as a kitten is playful, and it can be trained to make minor modifications in its natural behaviour in order to accommodate itself to the convenience of its human companions.

If the domestic cat has indeed retained to adulthood certain juvenile traits, the rest of the story falls into place rather neatly. Unlike his other domesticated species, man did not domesticate the cat. It domesticated itself. As humans began to farm cereals and to store grain – in the Near East – there must have been a large increase in the population of small rodents that feed on grain. They in turn would have provided food for cats. They would have been hunted by other carnivores too, but farming families would not have tolerated so readily some of the rivals of the cats – snakes, for example – that are more dangerous, although they did tolerate weasels. Throughout the world, and throughout history, people have always kept pets. Kittens would have seemed as attractive thousands of years ago to the boys and girls of the Near East as they appear to our children now. They would have played together, but as they grew into adult *libyca* cats, they would have become more difficult to handle. They would have reverted to their more natural role as ancient equivalents of our farm cats, but they would still have been regarded with some affection by the people who remembered them when they were kittens. Among all

Tom – very content to exploit a man-made environment

those cats there may have been some whose behavioural development was retarded. In some respects they remained kittens to the day they died. These cats would have become very tame, and so would have been kept throughout their lives by humans, and some of the kittens they produced would have inherited their 'personalities' to provide the nucleus of a breeding stock of inherently tame cats that in time supplied all pets cats. The protection afforded by humans would have enabled them to produce more kittens and to raise more kittens, mainly because they would have eaten better than they did in the wild. They would have been partly isolated from their wild relatives, and as they were taken into new regions when their owners migrated – which they did – their isolation would have become total, so that they began to evolve along a slightly different line, and so into a distinct species. Perhaps their small size is also a juvenile trait that accompanies their juvenile behaviour.

The fact that *catus*, the domestic cat, is a species distinct from its two closest wild relatives but not so distinct as to prevent interbreeding suggests that this is a new species which is still evolving and departed only recently from its ancestral stock. This, too, conforms to what we know of the cat. It is among the most recent of domestications, far more recent, for example, than the dog. Among mammals, it may be that only the Old World rabbit – originally from North Africa – and *Glis glis*, the edible dormouse, beloved of the Romans, and the Andean guinea pig have been domesticated more recently, although hamsters, gerbils and a few other small rodents are being domesticated now, as pets.

We do not know exactly when the cat was domesticated, but the first certain knowledge we have of it comes from a fragment of papyrus from the XVIII Dynasty of Ancient Egypt, which has been dated at about 1500 BC. It says, in part, 'This male cat is Ra himself, and he was called "Mau" because of the speech of the god Sa, who said concerning him: "He is like unto that which he hath made; therefore did the name of Ra become Mau".' The papyrus includes a picture of a cat holding a knife in one paw and the head of a python in the other, in the act of stabbing the python. If the cat is Ra, the snake is his enemy, Apep. The Egyptians revered cats, and *mau* was one of their names for it. Another was *kattos*. It seems unlikely that most Egyptians would have parted with cats to foreigners who did not share their religious beliefs, although the Egyptians were as partial to a little commercial profit as anyone, and it may be that the spread of the cat was slow. It is interesting, though, that like cattle, its origins are religious.

The early Greeks probably did not have cats. They controlled rodent pests with the help of weasels and ferrets, for which the Archaic Greek word was *gallê*, which described all the small members of the weasel family. Later, when they did have cats, they used the same word to describe them, and so it is not always easy to know to which species their writings refer. By about 300 BC, though, it is clear that *gallê* had come to mean 'cat', and in the fifth century BC Herodotus used the Egyptian word *kattos*. Aristophanes used *kattos* in 425 BC. Probably the Greeks regarded the cat as another kind of weasel, and Aristotle included the mongoose in the clan.

The spread of the cat through the Mediterranean region was most likely the work of the Phoenicians, who may have carried them on their ships, either by design or by accident. Apart from the Egyptians, other peoples had no special regard for this new recruit to pest control. The Greeks never coined a word for 'purr', and neither did the Romans, whose word *felis* meant cat but also, like the Greek *gallê*, weasel. No cat remains were found at Pompeii, and in an account of an infestation of rabbits that occurred in the Balearic Islands at about the time of the Vesuvius eruption, the animals used were ferrets. By the fourth century AD, though, the cat was well known throughout Roman Europe, and not only as a pest controller. From the time of Pericles, cat-skins had been used to make garments worn mainly by the poor, and the fashion reached its peak in the ninth century AD, when cat fur was valued as highly as otter fur. Charlemagne curbed the trade by imposing a price control on the skins of both species, and although there is no reason to suppose he did so in order to protect them, in fact the trade became less profitable and died for several centuries. By that time, the cat was firmly established in fact and in legend. An old Muslim tradition even held that the cat was created on Noah's Ark, when it was pointed out that the enterprise might fail because the rats would multiply and eat the grain stores. God caused the lion to sneeze and from its nostrils there leapt a cat, and the rats took cover.

Cats were among the species brought to Britain in Roman times, and no remains or authentic records provide evidence of cats in the British Isles before the occupation. It was European settlers who took cats to North America.

The Americans have no indigenous small cats comparable to *libyca* or *sylvestris*. Indeed, there were no cats of any kind in South America until about a million years ago, when the formation of the Panama land bridge enabled North American cats and sabre-toothed tigers to cross from the north. There are other small cats, of course. The ocelot (*F.pardalis*) measures up to four feet in length, not counting the tail, which is about a foot long, and it stands about eighteen inches high at the shoulder. Some ocelots have been tamed and kept as pets. The margay (*F.wiedii*) looks like an ocelot, but is about half its size, and the tiger cat (*F.tigrina*) is similarly small. Both species are shy forest dwellers of the tropics and not much is known about them. The jaguarundi (*F.yagouaroundi*), which is not closely related to the jaguar despite its name, is about half the size of an ocelot. It may be black, dark grey or chestnut in colour, with no patterned markings in its fur, and, unlike Pickle, who was worried by her encounter with a slow-worm, it kills and eats snakes as well as mammals up to the size of a monkey and fish. Even the largest of the American cats, the jaguar (*Panthera onca*), is only about four feet long from its nose to the root of its tail, and has short legs, but it is heavily built and looks larger. All these species are quite different from the domestic cat and geographically they are confined to the American tropics and subtropics, ranging as far north as Texas. The northern cats are lynxes, and evolutionarily even more remote from *libyca* and *catus*.

It is possible to trace the migrations of domestic cats from the Mediterranean to every corner of the world, because unlike any other domesticated species it has always been uncommon for cats to be bred and sold, and so transported deliberately into new areas. The pedigree Persian, Siamese or Abyssinian is bred for its appearance, and you will have to pay for a pedigree kitten just as you would pay for any pedigree animal, but the great majority of cats are still not bought and sold. They are too plentiful to have a market value. Since they are not transported over long distances by their human companions – or not usually – most cats live their entire lives within a few miles of the place where they were born. Thus local populations can emerge that possess certain recognisable features, and by noting the frequency of these features in different places it is possible to discover where they originated, how they have spread, and in some cases even to measure the rate of spread.

The most obvious feature is the pattern of the fur. This is determined genetically, but there may be as many as forty genes involved and so many variations are possible. We know the original colour of the first domesticated cats, for they must have resembled the modern *libyca* cats. At some point in their evolution they acquired a gene that modified their overall brownish colour by providing it with dark spots, and this gene mutated to provide dark stripes. Probably later another gene was added which prevented some of the fur from developing any pigmentation at all, so it could produce white patches. Other genes turned the dark spots and stripes to a red colour and little by little, minor mutation by minor mutation, the entire range of colour and pattern became possible. Some of these genes are recessive – such as that which confers a pale fur colour on Siamese – and others are dominant – such as that for white patches. It was J. B. S. Haldane who first suggested that such features should be used to plot the gene frequency of domestic cats throughout the world, and a project to do this began in London in 1949. Since then it has been continued into several other European countries, a few countries in North Africa and at the eastern end of the Mediterranean, Australia and New Zealand, Thailand and, in the Americas, the United States, Mexico and Brazil.

In the seventeeth century, a version of the traditional tabby pattern, in which the stripes are broad and twist into whorls, became very popular in England, and some of these cats were taken to North America. They were taken mainly to Boston. We can deduce this because today 42 per cent of Boston cats are of this pattern. As the toms sought females in neighbouring ranges and so widened the distribution of their genes, the pattern spread. In San Francisco and Dallas 27 per cent of cats bear these markings, in Houston 29 per cent, and in

Mexico City 23 per cent. The English settlers who moved west often took their cats with them, so there are high populations of cats with these markings in the cities they founded. To the south, most cats are descended from Spanish ancestors and this particular pattern of fur marking never became popular in Spain. In the Old World, the same pattern can be traced as it moved eastward. 81 per cent of London cats bear it, and the frequency falls steadily as you cross continental Europe, until it reaches 15 per cent in Ankara. Polydactylism – the possession of six or more digits – is also determined genetically, and its migrations can be traced, too. In the United States it seems to have appeared first in Boston, probably early in the eighteenth century. Since then it has been spreading at its natural rate, and has reached as far as New York. The fact that it has taken nearly three centuries to travel a mere two hundred miles shows the slow rate at which genetic information spreads – in this case about a thousand yards a year. There is a high population of six-toed cats in some parts of the Canadian maritime provinces, however, and this can be explained by supposing that families loyal to the British fled from Boston during the revolution, or just after it, and took their cats with them.

Cats that moved into Asia also continued to evolve, and some of them acquired features that Europeans and Americans found very attractive and exotic when first they saw them. The Siamese – one of the ancestors of the Burmese – had collected genes that gave it blue eyes and fur that developed dark pigmentation only when the skin temperature was maintained at less than 98 degrees Fahrenheit. This was discovered half a century ago by two Russian scientists, N. A. and V. N. Iljin, who shaved away the fur from dark areas on Siamese cats, covered the naked skin with warm cloths, and the cats grew pale hair. It is possible that the pale Siamese fur is a kind of partial albinism, because Siamese cats share with many albino animals a faulty arrangement of their optic nerves. The nerves that transmit impulses from the retina of the eye lead to the lateral geniculate nucleus, which is a region of the mid-brain. It consists of two layers, upper and lower, and each nerve from one eye enters the upper layer and its point of entrance is matched precisely in the lower layer by the entrance of a corresponding nerve from the other eye. The result is that the brain receives two images, one superimposed on the other. Both hemispheres of the brain are supplied with an equal amount of information, because some nerves from each eye cross to the opposite hemisphere, while most connect in the same hemisphere as the eye itself. In the Siamese, though, an abnormally large number of nerves cross to the opposite hemisphere, and they are paired incorrectly at the lateral geniculate nucleus, so that the images that are superimposed on one another are not identical. The neural message reaches the brain in a garbled form. The cats cope with this – and can see perfectly adequately – either by suppressing the discordant information or, depending on the location at which the nerves enter the lateral geniculate nucleus incorrectly, by re-routing the nerve impulses to remedy the fault. This strategy causes their eyes to cross, for reasons that are not understood.

Asian cats also developed a gene that gave those in which it is dominant a deformity in the tail. This is best known as the kink which is found sometimes in the end of the tail of a Siamese, but it can take other forms, and any domestic cat with an unusual tail – or no tail at all, like the Manx – is probably descended from Asian ancestors.

Pickle had caught her vole and trotted off with it, her head held high, tail erect. She caught a glimpse of Peter, but ignored him. He was unimportant. Like all animals, she exploited her immediate environment as best she could, to obtain whatever she might need, but unlike most animals, she and her kind had learned to regard humans as a part of that environment and to exploit them, too. She was fairly independent of them, but her many close relatives who at that very moment lay curled comfortably on sunny windowsills had come to rely on them. The trick had been to control their fear, to learn to tolerate handling and even to enjoy it, as the necessary price that had to be paid for an easy life. In a way they had to prostitute themselves, but once they had mastered the art, the rewards were impressive. They allowed the domestic cat to live longer and more healthily than its wild ancestors, and it managed not to sacrifice its independence in any real sense. It exploited the humans, but was not unduly exploited itself. If humans chose to believe their pets were domesticated, that was up to them. It was and is true, up to a point. By a process of selection, in which we have played a part, cats have modified their behaviour to make themselves more attractive to people, but to demand servility of them would be to demand too much. So far as cats are concerned, humans are the providers of food and of habitats into which they can fit themselves and that is all. Pickle's hedge, the barn, the garden, the armchair by the fire, differ only in the amount of food and degree of comfort they provide and like all domestic cats Pickle was quite capable of exploiting a new environment if the benefits looked tempting. It is the habitat that matters and humans are the burden cats must bear to gain access to it. If the habitat could be maintained without them, few cats would mourn the passing of humankind.

Don't misunderstand this. Cats are no different from dogs in this respect. Dogs have developed different forms of behaviour which lead them to exploit human environments in rather different ways, but they are exploiting people no less than are cats, and they, too, would be content with a man-made environment from which people had departed. We do not exploit our pets to anything like the extent to which they exploit us. Has any pet owner ever doubted it?

Pickle walked into the barn, dropped her vole on the floor, and growled at the other cats to keep them away as she ate it, head first. In a few moments it was gone, and she greeted her friends before sitting on a hay bale to enjoy a thorough wash.

The next generation

AT THE END of March, Peter had noted that all three females appeared to be pregnant. There was no outward, physical sign of pregnancy at that stage, and his observation was based entirely on changes in their behaviour. Each of them had become sexually receptive every fifteen days. When Tom appeared, they would make their 'chirruping' calls, rub themselves against him, roll on the floor, and perform a seductive display that often ended in copulation. It was that activity that ceased, quite abruptly, and the four animals seemed to lose all interest in sex. There was only one explanation. The period of receptivity had served its purpose. Implantation had triggered hormonal changes in the bodies of the females, so removing their urge to mate, and as they changed so did Tom.

Tomcats do not quite deserve their voracious sexual reputations. True, like all male cats, Tom sought to spread his genes as widely as he could through the local feline population, and to this end his search for receptive females never ended, but the females had to be receptive. It was always the female who made the first advance by indicating that she was in oestrus. Unless she invited sex, Tom would not force his attentions upon her.

Cats reach sexual maturity anywhere between five and nine months from the time they are born if they are female. Males may take up to eighteen months. Just how long it takes depends on several things, but the most important factor, almost certainly, is the quality of the diet. House cats, which are usually better fed than their more self-reliant cousins, usually mature earlier. It is also possible that the maturation of males may be delayed where a mature and dominant male is present.

There is more to it than just diet, though. The season of the year also plays a part. Rudi van Aarde, a young South African zoologist, studied a colony of cats on Marion Island, in the Indian Ocean, about 1,300 miles south-south-east of Capetown. The ancestors of these cats had been taken to the island during the nineteenth century by whalers, sealers and meteorologists who had visited at various times. Van Aarde found that mating occurred mainly between July and January (remember the seasons are reversed in the southern hemisphere) and litters were born between September and March. It was only between July and April that pregnant and lactating females were found. The female kittens matured in the first breeding season after they were born.

The advantage of this arrangement to the cats is obvious. The kittens are born at the time of year when food is most plentiful, not just for them but also for their mothers. Probably it was the changing length of day that triggered the hormonal activity that governed mating, but it could be that the pattern evolved naturally. Since kittens born in autumn or winter would be less likely to survive than those born in spring or summer, after a few generations most young cats would be having their first litters at the same time of year, because they would have been born at the same time from parents who were also born at that time of year.

Marion Island is a grim place. It rains there on 300 days in the year, the westerly winds blow almost incessantly, and it is cold. Animals living in such a harsh environment are likely to develop a more markedly seasonal breeding pattern than those living amid friendlier surroundings.

It was not just the females whose sexual behaviour was controlled by the seasons. The males also underwent physical changes that made them sexually more active at one time of year than at others.

The average litter size on Marion Island was 4.63 kittens, and 42 per cent of them died before they were weaned. This meant that in spite of the harsh conditions the population was increasing fairly rapidly. Indeed, it was the population explosion of cats on the island that led to the study, since they were destroying much of the local fauna, which consisted mainly of sea birds that were being subjected to predation from hungry cats. The feral cats of Marion Island were highly successful, and van Aarde estimated that their average age at death was about eight years.

Pets often live for much longer than this, of course, and on average they can continue to produce litters for eight to ten years. It is rare for a female cat to give birth to kittens if she is more than fourteen years old. Her reproductive life ends gently, though, with no great drama. An increasing number of copulations prove sterile, and the number of kittens in each litter tends to become smaller, until eventually she ceases to be sexually active. While she remains active she is capable of producing litters up to four times a year – provided the conditions are ideal. The Marion Island cats produced an average of two litters a year each. As the study of the farm cats was to suggest, the number of young that are born often relates directly to the number the environment can support, and the size of a population is not regulated solely by infant mortality.

As anyone will know who has tried to live with an unspayed female Siamese, a cat in oestrus cannot be ignored. In our latitude she will come into oestrus for periods of about five days at about fifteen-day intervals during the first six months of the year, but in some individuals the cycles can continue from Christmas to the middle of autumn. As her hormonal balance changes, each cycle develops through definite phases, with characteristic behaviour in each, until it reaches its peak. At this point she makes incessant demands. Head down, rump raised, she will tread with her hind legs, calling all the time in a deep voice. There is only one way she can be subdued, and if she is confined indoors with no male on hand, she will continue to demand until the cycle reaches the point at which her hormone balance changes again and it is all over – for about a fortnight! Because her voice is deeper and louder in any case, the calling of a Siamese at these times can be deafening.

Pickle, Smudge and Domino were pregnant, then. Or were they? Early in April, Peter began to wonder about Smudge. She looked pregnant by this time, but he saw her seduce Tom in the field beside the road. There were many attempts at copulation, and although he saw none that looked as though they were successful, Smudge did behave as though she were fully receptive. Peter was puzzled by her behaviour. What was the point of sex if the female was pregnant already? 'An alternative is that it all has some social function,' he wrote. The mystery deepened when, three days later, he saw the two cats trying to copulate again. He commented: 'Whether or not she is pregnant is still doubtful. She looks fatter than normal, but not to the same extent as Pickle, and there is no swelling of the mammary glands, which should have started by now.' Peter assumed for the time being that Smudge was not pregnant and turned his attention to Pickle and Domino.

As the pregnancies advanced, life among the cats assumed a peaceful, sociable pattern. Like all sensible expectant mothers, the females ate well and rested much. Tom and Domino became more

Tom mating with Smudge as Domino looks on

friendly towards one another than they had been, and Tom tried to mount her once or twice, but was rebuffed. She developed a habit of kneading, as a means of communicating something to Tom. Peter had seen her knead during cold weather, when she wanted to ease her way into a heap of sleeping cats for warmth, but he had no idea why she was doing it now. Tom took no notice of it.

Life was easy. When the sun shone the cats sunbathed a good deal. Sometimes they would go off hunting. They did not like the rain, and would return to the barn, even from a hunting trip, if the weather deteriorated.

Gestation in cats takes from 63 to 65 days, and by the end of April both Pickle and Domino were very swollen and their nipples had become prominent and pink. Before she gives birth a cat makes a nest. Cat owners, knowing this, usually try to provide a nest in some suitably warm, secluded place. Shoe boxes in airing cupboards seem popular, but wherever the nest is placed it must be lined with cloth or paper. In most cases the cat will accept this, adapting it to her needs by clawing and sometimes biting the lining to tear it into the shape she prefers and by rubbing herself against the sides and bottom. She moults shortly before giving birth, and her rubbing detaches loose hairs which become part of the nest lining. She will spend periods of ten to thirty minutes in nest building, alternating these with sleeping in the nest, except when she needs to eat or to go outside. By the last week in April, though, neither Pickle nor Domino had shown any sign of looking for a suitable nesting site. Smudge continued to behave enigmatically. Peter did not see her looking for a nesting-hole either, but she did go off by herself rather often and he had no idea of the purpose of her journeys.

The scientists and film-makers had their own ideas about what should happen next, for they planned to observe the births and to film them. The technique for doing this is well established. A snug cavity had been made among the hay bales, with an entrance that might make it appealing to a cat. It was fitted with light bulbs wired to a rheostat, so they could be faded up slowly. This gradual increase in illumination would not disturb a cat in the least. Although this den appeared solid, one side of it contained a plate glass window and, behind the window, a larger area in which Maurice could erect his camera and tripod and operate the rheostat, and Peter could observe and make notes.

The trouble with filming animals is that they do not respond to direction, they fail to rehearse, and their performances are unpredictable. Peter, by this time almost a nocturnal animal himself, had been up all night with the cats, and was sleeping in his hut when Pickle decided to produce her kittens. What was worse, she chose to ignore the nest that had been so carefully prepared for her, and gave birth outside it. To cap everything, her kittens were born during the lunch hour. Peter's rather terse note is commentary enough:

Pickle had three kittens between 1230 and 1330 on 6.5.78. The birth took place on the floor of the barn, just underneath the entrance to the special hole, and the kittens have remained there despite close approach by the farm cats and chickens. The farm cats pay no special attention to the kittens and Pickle allows them to approach within sniffing distance. Pickle occasionally spits at the chickens, instead of running away as usual.

In fact, her reaction to the chickens showed considerable courage. A farmyard chicken may seem harmless enough to a human, but it can be highly aggressive towards any animal smaller than itself, and most cats leave chickens strictly alone. The slightest interference can make the bird charge, screeching and flapping its wings, and a cat cannot cope with it. Pickle had little choice, of course, for the chickens might well have killed her kittens if she had not stood her ground.

By the time Peter and Maurice arrived on the scene, all three kittens had been born and Pickle was 'clearing up'. We can only guess at what happened, but our guess is a reasonable one, because the birth of kittens has been observed many times.

As the contractions began, Pickle would have sought a quiet place in which to lie down. Of course, she should have gone to her prepared nest, but perhaps she left matters a little too late, so that her kittens were born in the feline equivalent of the back of the ambulance, or perhaps her inexperience caused her to be inadequately prepared. We are used to thinking that all animals except ourselves are guided by what we like to call 'instinct', so that in the important matters of life they require no instruction. Among many animals this is more or less true, but the more highly evolved species, such as the cat, need to be taught much of what they must know, and even in such basic matters as reproduction an 'uneducated' cat may get it wrong, just as an uninformed human mother may. After all, a pregnant woman will give birth, whether or not she knows what is happening to her and whether or not anyone is there to help. Although cats are more independent at this time than humans, and their genetic 'programming' and physiology serve them better in some respects, their programming is not always perfect.

As she lay awaiting her first kitten, Pickle would have licked herself more or less continually, especially in the genital region, and as the first kitten appeared her licking would have been transferred to it. This licking undoubtedly had a hygienic effect, providing a mildly antiseptic environment through which the kitten must pass. In some way it also eased the birth, and Pickle found it soothing. As they emerged, the kittens had to be licked, roughly and vigorously. They are born shrouded in a membrane. This must be removed, their mouths and nostrils must be cleaned to ensure that the airway is free from any obstruction, and the new arrival must be made to start breathing. Her licking was the equivalent of the wash and slap with which the newborn human baby is greeted, and it served precisely the same purpose. She would have eaten the membrane, and when the last of her kittens was breathing and the placenta emerged she would have eaten that, too, including the umbilical cords down to within about an inch of each kitten. At this stage it is not unknown for a cat to continue eating and to eat her kittens – occasionally all of them. In Pickle's case, though, all went well and her kittens were born alive and healthy.

When Peter found her, it was clear that Pickle had not been unduly distressed by the births. The fact that all her kittens were born in less than an hour shows that there were no complications. The

Pickle with her first litter

intervals between kittens during the birth of a litter can vary widely. Sometimes one kitten follows another in less than a minute, but the usual interval is more like fifteen minutes, and it can be as much as twelve hours.

Almost certainly this was Pickle's first litter, and her inexperience made her a far from perfect mother. Usually a cat will stay with her kittens all the time for the first two days. During this time she neither eats nor drinks and she does not urinate or defecate. After two days the kittens have advanced sufficiently for her to be able to leave them briefly to attend to her own needs, but even then she cannot afford to be away from them for long. They need the warmth of her body and they need to suckle frequently. Peter recorded that between 1230 and midnight on the day they were born Pickle left the kittens twice, for periods of two to three minutes. The next day the absences grew longer and more frequent. On once occasion she left the barn and went almost as far as the church. When she was nearby, Pickle would be attracted by the cries the kittens made because they were cold, but while she was with them she would make no attempt to retrieve a kitten that crawled away from her and felt lost. It had to find its own way back to her.

The following day, 8 May, Pickle moved her kittens into the nest in which Maurice and Peter had hoped they would be born. Young kittens are usually moved in this way when they are a few days old. Probably it is a security measure. During their first few days the kittens cannot be moved far, but little by little the smell of them and the noise they make may attract predators who are able to pinpoint their precise location and bide their time until the mother has to leave them unprotected. By moving them to another secure site she avoids detection for longer. It is not difficult to see how such behaviour may have evolved. There is a period of about four weeks during which the kittens are unable to move far from the nest, but their mother must find food for herself and, as they begin to take solids, for them. They have to be left alone. If moving them to a new site and greater security is behaviour that is controlled, and so transmitted genetically, the cats which possessed this instruction would produce offspring that survived more frequently than those which lacked it, so that eventually all cats would possess this necessary item of programmed behaviour.

The kittens must be carried to their new home, since they are unable to make their own way there. The correct way to carry kittens – if you are their mother – is by the loose skin at the back of their necks.

They must be carried one at a time, obviously, and it was by watching cats move their young kittens that scientists acquired a curious item of information. The mother carries a kitten to the new nest, drops it with its fellows, and returns for another kitten. After she has carried the final kitten she returns to the now empty nest and takes a last, careful look round it. What is the purpose of this visit? The nest is empty. It contains nothing that could conceivably be of interest or value to the family, and the mother does not interfere with the old nest in any way. The only possible explanation is that she is looking for more kittens, and this suggests that cats cannot count. They have no concept of number, and a female does not know how many kittens she has.

Pickle's kittens were grey and ginger, white, and black and white. The transport of kittens to the new nest is a hazardous business. While she carried the white one into the secure hole she had found, Pickle left the other two outside. Then she carried in the grey and ginger one. The white one died on the evening of the 8th, probably from being left alone for so long. At least, that was what Peter thought at the time, and he did not consider the death a serious setback for Pickle because the remaining two kittens seemed healthy enough.

Once they were secure in the new nest, Pickle was able to spend more time away from them, hunting. But she had to spend much of her time with the kittens and so she took little part in the social life of the cats. Combined with Tom's prolonged absences, this greatly reduced the feline activity in the barn, and when he was there Tom was inclined to be irritable with Domino. It was a quiet time, and attention now turned to Domino.

Having missed the birth of Pickle's kittens, which continued to grow sturdily, Peter was determined not to miss Domino's confinement. He and David had calculated when her kittens would be due. It was her first litter, too, and Peter was under some strain, which was reflected in his notes.

> As of 2300 today, Domino has not had her kittens. I am keeping her under close observation with checks about once an hour for as long as I can stay awake. The kittens have been moving so strongly that the movements are visible externally. This began about 2300 on 19.5.78 and at the time I thought it was a signal for their imminent emergence. Domino spends most of her time wandering apparently aimlessly or just sitting. She has apparently still not found a hole for the kittens.

She was not his only charge. A large question mark still hung over Smudge.

> Smudge, if she is pregnant this time, is due to have kittens between the 9th and 14th of next month (June). Once Domino has had hers, I should just about have time for some before-birth spatial data on Smudge, which would give us a before, during and after pattern . . .

Poor Peter had a long time to wait, because Domino's kittens were born several days later. Shortly before the birth she moved into the nest in which Pickle was still tending her kittens. She produced five kittens, four of them black and white and one grey and white.

What Peter and Maurice witnessed through the glass of the den turned out to be one of the unexpected highlights of the film and a key event in the scientific study. The event was filmed, but it was much more than the birth of a litter of kittens. Pickle acted as a midwife to Domino. As the kittens were born, it was Pickle who licked away the birth membranes and cleaned Domino's fur. Peter noted that 'Domino seemed incapable of handling the birth alone and at times lay immobile while Pickle tended the kittens. Neither cat showed the slightest sign of aggression towards the other or towards any of the kittens. As far as I could tell, the cats and kittens did not discriminate in favour of their own families.'

The film-makers were highly excited about the filming of the births and the co-operation between the two mothers. The two biologists were not especially interested in the births themselves, but they were fascinated by the co-operation. This was something that was suspected of happening – and many viewers wrote later to report their own observations of similar behaviour – but that had never been observed between cats whose social organisation had been so thoroughly studied. An unedited, mute, copy of this piece of the film was sent to an international conference on cat behaviour that was being held at Syracuse, and David Macdonald and the producer Peter Crawford showed it to the assembled scientists. At that stage the film was technically crude, but the scientists watched spellbound and it gave rise to much discussion.

The argument in favour of the 'selfish gene' hypothesis is persuasive. It holds that from an evolutionary point of view what matters is the survival of genetic material, which is transmitted from generation to generation and which lasts much longer than any of the individuals that hold it at any time. In a sense, animals (and plants) are 'survival machines' for the perpetuation of genes, and therefore all individuals are inherently selfish. They will do what is necessary to pass on their genes and ensure their survival. The hypothesis is subtle, however, for it predicts behaviour that may appear to be altruistic. When Pickle helped Domino, for example, it looked as though she was devoting time and trouble that she could have spent hunting or caring for her own kittens in order to assist into the world kittens that might grow into hunting and breeding rivals for her own. How can this be explained? Because Pickle and Domino were full sisters, half the genes each of them carried were also carried by the other. Since Tom was the father of both mothers and both litters – at which point the description of the familial relationships strains the English language to breaking point – the kittens had a great deal of genetic material in common. Therefore it was almost as satisfactory for Pickle's genes to ensure the survival of Domino's kittens as it was to ensure the survival of Pickle's. Because Tom was simultaneously grandfather and father of the kittens, these newest offspring were related among themselves even more closely than were Pickle and Domino to one another.

Pickle and Domino were probably daughters of Tom and Smudge. Genetically, therefore, each of them was half Tom and half Smudge. They donated half of their own gene complement to each of their kittens which were, therefore, one-quarter Tom and one-quarter Smudge on their mother's side. Tom, the father of the kittens, then contributed the remaining half of their genes, so that each kitten was three-quarters Tom. It is hardly surprising, then, that of the eight kittens produced, six were black and white – like Tom! Nor is it surprising that the two litters should have been run together. These kittens were much more than mere first cousins! By helping her sister, Pickle was helping herself and her genes.

The delight and enthusiasm of those who watched these events were reflected in the commentary to the film, and may have suggested to some viewers that such co-operation was unknown. This was not the intention. The interest lay in the fact that this was the first time it had been observed in the context of a colony whose relationships were so well known and where the significance of the event could be scientifically interpreted. It was also the first time that such co-operation had been recorded on film.

Most of the examples of similar co-operation that were described by viewers involved cats that were closely related. Some described such co-operation between two cats, one of which was a neutered male. 'On Wednesday of this week,' one viewer wrote, 'my husband and myself witnessed an even more amazing phenomenon when one of our three neutered tomcats acted as midwife to the female during the birth of her one kitten, at her own insistence.'

As though that was not curious enough, several viewers described co-operation between cats and dogs, not during the birth it is true, but in the rearing of the young. 'When I was small,' one lady wrote, 'our cat helped the Airedale bitch look after her pups. The bitch returned the compliment when it was the cat's turn to be mother. They each carried the other's babies back into the house when they strayed.' A Belfast viewer had a cat whose chosen site for the secure nest into which to move her young kittens was the basket owned and occupied by her friend, who happened to be a Dalmatian bitch. She placed the kittens in the basket, apparently to the distress of the Dalmatian, who left the 'nest' to the cats, but who returned later and assisted in watching over the kittens. A lady who reported co-operation between two pedigree, and quite unrelated cats, went on to say that the maternal one of the pair 'took a five-week-old Jack Russell puppy under her wing . . . washing it and mothering it just as if it were a kitten'.

We cannot interpret such reports or find explanations for them. If it were possible to watch them closely, it might be possible to discover why they happened. The clues we would seek would be found in each tiny movement, gesture, facial expression and sound made by the animals involved. This is the 'language' by which they communicate, and a main purpose of all studies of behaviour is to try to translate it. This is why Peter recorded the actions of the cats in such minute detail. It is very probable that prolonged – often lifetime – association with humans and isolation from their own social structures cause large distortions in the behaviour of domestic and pet animals that amount to confusions concerning their own identity.

This is not to say we disbelieve the reports. It may well be that the world contains wonders more magnificent than those we have seen for ourselves, but our own modest wonders impress us simply because we have seen them, and every small detail of them is recorded. In his narrative to the film David said, 'As Domino chewed the umbilical cord to her first-born kitten, Pickle leaned over to sever it. As the second kitten was born, Pickle licked both it and its mother, and again severed the cord.' It was indeed a remarkable event.

But the euphoria was over, abruptly. Some of the kittens caught 'cat flu' and within a week all the members of both litters were dead. Everyone was stunned. The disease progressed with devastating speed, and no veterinary treatment could have saved them. Peter buried the kittens behind the barn.

Pickle and Domino appeared quite undisturbed and quickly reverted to their former patterns of behaviour, but the humans were distressed. As Peter and David struggled with the benefit of hindsight to reconcile themselves to the bereavement, all kinds of questions came to mind. The same questions occurred to many viewers, who said the adult cats should have been vaccinated to protect them and their offspring. The sad fact is that there was absolutely nothing anyone could have done to anticipate the deaths or to prevent them.

'Cat flu' is a generic term that is used to describe two related diseases. It used to describe feline enteritis as well, but this is now considered separately. All three complaints can kill, and all are caused by viral infections. Feline enteritis can be prevented by vaccination, and the vaccine is effective. Both forms of flu can be controlled, too, by vaccination with a vaccine that is effective for about twelve months. However, cats less than twelve weeks old cannot be vaccinated because their bodies have not developed sufficiently for them to be able to manufacture antibodies, and although adult cats can be vaccinated, the immunity vaccination confers on them is not communicated to their offspring, which remain vulnerable. There is some confusion about this, because the vaccine that prevents distemper in dogs does provide an immunity that is transferred to their offspring, protecting them for about twelve weeks.

The two forms of flu produce different symptoms, so it is quite easy to distinguish one from the other. In one form, the victim is obviously unwell, depressed, apathetic, loses its appetite and suffers from running nose and eyes, sneezing a good deal – very much like a human with a very bad cold. The other form causes a cough and ulceration around the nose and mouth. In both cases the infection attacks the upper respiratory passages, and although it is most unpleasant, and often causes fever, it is not especially dangerous in itself. The danger arises secondarily, from its effects on the cat's sensory system.

Cats depend on their senses of smell and taste much more than we do. It is by means of these senses that they decide whether or not an item they meet constitutes wholesome food. The ability to distinguish between good meat and bad is of obvious importance to a carnivore – bad meat can be as dangerous as it is to us – and the need to distinguish is ingrained so deeply that a cat will take no chances. If it is unsure, it will not eat. We know only too well that a bad cold deprives us of our senses of smell and taste. To us this is merely inconvenient. To a cat it is crippling. Perhaps the animal might be prepared to be brave and take a risk were it not for the fact that its blocked nasal passages affect it more severely than we can imagine. This is because the passages themselves are relatively narrower than those in a human, so they block more easily. The suffering cat has real difficulty in breathing, and added to the loss of its most important senses and its general distress this can produce a loss of appetite that is total. The animal cannot tell whether the food in front of it is safe to eat or not, and in any case it has not the slightest wish to eat. Victims of cat flu actually die from starvation. Once a cat becomes infected, the only possible cure lies in good nursing and if necessary forcing the patient to take nourishment.

An adult cat may well survive the disease, and it is so common that we may assume that many adult cats that live independently of humans have had flu and recovered. This is not the end of the story, however, because their recovery will have left them carrying antibodies to the virus that provide them with protection against future attacks. Should they encounter the virus again, they will exhibit no symptoms and will feel perfectly well, but nevertheless they will be infectious and able to pass on their infection to cats that do not share their immunity. This immunity can be lost at any time, but while it remains they are symptomless carriers. It is quite possible that Tom and perhaps some of the females, too, were symptomless carriers of the virus. Peter had noted that Domino seemed listless, thin, and without appetite. It is quite possible that lacking immunity herself she had caught the disease, probably from one of the other cats that was immune and so displayed no symptoms. She passed it on to the kittens – or the carrier did. Domino recovered, but the kittens did not. A kitten lacks the reserves of strength that are needed to combat so virulent an attack, and unless it is fed at frequent intervals it can starve to death very quickly.

Whatever the cause, the tragedy could not have been avoided. Within a week Domino was sexually receptive again. Pickle was especially playful, and all four cats seemed to be settling back into the way of life they had enjoyed before the kittens were born.

Smudge, though, behaved oddly. She disappeared from the barn often and for long periods. Peter could tell from the signal he received from her radio collar that she was not moving far, and he did not try to follow her. During the first half of June the weather was very wet and the other cats spent much of their time in the barn itself, with Peter watching them and pausing every hour or so to check on Smudge's movements. In the middle of the month, Peter was alarmed by the fact that for more than two days Smudge's radio signal had come from a constant direction. Perhaps she was trapped or injured, perhaps ill – or worse. Peter decided he must find her.

Beside the barn there were various other sheds and outbuildings that Maurice had not used for some time. The jungle of metallic objects reflected the radio signal, and made the search for Smudge more difficult, but at last Peter tracked her down. In a dark corner of the old cow byre he found her curled up in a ball. As he approached, fearing the worst, Smudge lifted her head. She was purring and Peter could see that she was not alone. Nestled against her neck was a kitten. Smudge's absence was explained. Without warning, she had withdrawn from the group and produced a single offspring – a ginger and white male.

It was much later, when there was time to reflect on what had happened, that a pattern emerged. The adult cats were good, successful hunters, but the ranges they had established around the farm probably produced little in the way of surplus food. It is doubtful whether the habitat could have produced sufficient small mammals and birds to support all the kittens from the earlier large litters, if they had survived the ravages of cat flu. Unless Peter had intervened to feed them, most probably some would have starved. One kitten, though, could be supported without difficulty, and it was curious that it should

Smudge secretly gave birth to one male kitten : Peter named him Lucky

have been Smudge, the oldest and most experienced of the females, that spent so much time seeking a suitable nest site which was isolated from the other cats. Her kitten was born after the other litters, but even if the timing had coincided, he would have had a greater chance of avoiding the disease in this nest on his own.

Was it all coincidence, or was Smudge behaving differently from the others in order to improve the chance of survival for her kitten? Why did she produce only one? It is not unknown for the number of young born to a group of animals to be conditioned by the amount of food that is available. Maybe the change in Smudge's behaviour and her own internal physiology was in response to her immediate environment.

In the weeks and months ahead everyone's attention focussed on the new kitten. The cats cared for him, fed him and educated him, while Peter and Maurice watched and filmed his progress. The unexpected arrival of this new member of the Devon colony brought a whole new tempo to the lives of the cats and their human observers. It soon became apparent that the new kitten had every chance of surviving. Peter gave him a name. He called him 'Lucky'.

Learning to be a cat

LUCKY WAS very small, blind and deaf. No more than a tiny feeding machine he was quite helpless. At first he could not be left alone for more than a few minutes without becoming distressed.

Kittens are born with both eyes and external ears closed, and even after they open it is some time before the organs can be used in a controlled fashion. They do seem to have a sense of smell, though, and they probably use this to find their way around the nest.

Although he was so helpless, he was anything but immobile. His legs gave him some support, at least when he was hungry. After he had fed, his belly swelled so much that his paws no longer reached the ground, and as the swelling subsided, as it did between meals, there was a period during which his weight remained too great for his limbs to lift. His mode of travel had nothing to do with walking. Like a human baby, he crawled.

His paws helped in this because his claws could not be retracted and they tended to catch in any irregularity of the surface and so give him a grip. They did not move him by themselves, even so. By contracting the muscles on one side of his body he would make his whole body curve. Then, a few moments later, he would relax those muscles and contract the ones on the other side, so his body curved the other way, each time leading with his head. This alternate swaying motion sometimes continued for quite a long time. All kittens do it, and it seems likely that it is automatic. Had he been on a smooth surface, it would have carried him nowhere. It would have been nothing more than a long, slow, writhing kind of movement. On an uneven surface, however, it enabled him to get about, sometimes by travelling in a circle admittedly, but more usually towards Smudge. As often as not it was she who initiated the movement, by licking him. When she licked on one side of his spine, that side would contract, exposing the other side. She would lick that side and the automatic swaying would begin. As he moved, his legs usually dragged behind him until they were a little stretched, when he would bend them up, very quickly. Like all good cat mothers, Smudge began her washing of him at his head, and this made him move towards her. He was quite capable of moving away, though, should she treat him roughly.

After a couple of days, when he could be left for short periods, Smudge began to make hunting trips for food, but her return to the nest always meant a thorough wash for Lucky. Smudge would nudge him with her nose, purring the while, and then begin to wash his head. When that had received adequate treatment, she would move on to his back and sides and then she would tip him over so he tumbled on to his back. She washed his belly, licking away any urine or faeces, but when at last she was done, and Lucky was sparkling clean again, she seemed to lose interest. She never remembered to turn him back on to his feet. He had to do that for himself. It was his first lesson in independence.

When he was hungry, which was often, he would wriggle his blind way towards Smudge until he made contact with her. Because of the wriggling movement that took him to her, he arrived with his head still waving from side to side. As he pushed his chin against her flank the swaying movements became smaller, but they helped him to find a teat in the small depression his head had made in her body. He never pushed his head beneath her body, and always pushed with his chin, never with any other part of his head. He did not have to be taught to keep his nostrils clear.

It was June, and mild, and there was no risk of him freezing to death. Indeed, it seems that kittens need the warmth of their mothers much less than is often supposed. Some years ago scientists placed kittens one at a time between two hot water bottles. The bottles were arranged so that as it swayed each kitten touched first one bottle and then the other. One bottle contained water at room temperature, the other water at the temperature of the adult cat's fur. The kittens showed no preference for one bottle or the other. That led the investigators to find out whether it was the texture of the mother's fur that provided the attraction. They filled both bottles with water at the same temperature, but wrapped one in fur and the other in cloth. Again the kittens did not prefer one to the other. The experiment was repeated yet again, only this time the bottles were both wrapped in cloths, one of which was clean and recently washed, while the other was a cloth taken from the nest where the kittens and their mother had been lying. There was still no preference.

Lucky looked cute, but clearly his awareness of his surroundings was very limited, if indeed it existed at all. He was nothing but a small feeding machine. His utter helplessness did not last for long. On 19th June Peter found that both eyes were open, and checked to confirm that they seemed to be healthy. The eyes of kittens open after seven to ten days in most cases, and the outer ears begin to open at about the same time. It is not until the kitten has been alive for about a fortnight, however, that its eyes will follow a moving object, and it is about then that it begins to respond to noises. It twitches its ears at sounds and mews in response to its mother's calls.

From the very beginning, sound is more important than sight. By the time Lucky was a little over three weeks old he could turn his ears very quickly to direct them towards the source of a noise, and loud noises alarmed him. If he were watching something that moved – he was capable of this – a sharp noise could distract him. He could see as well as he could hear at this stage, both senses developing at the same rate. At a fortnight old he could move his paws close to Smudge's nose while she was licking him, and a week later he could follow with his eyes objects up to a foot away from him, provided they moved slowly. If they moved fast he would lose them. After that his eyes were entirely functional, and further development depended not on the eyeballs themselves but on the strengthening of the muscles that controlled them. His ears were not fully formed until he was more than five weeks old, and then, because his skull was so much smaller than that of an adult, his ears looked very large.

As soon as Lucky opened his eyes, Pickle began to visit. She spent a good deal of time with Smudge and sometimes she was left alone with the kitten. One of the first sights on which young Lucky may have tried to focus his eyes may well have been a play-fight between his mother and the 'aunt' who was also his elder sister. Not being refined lap cats, the games were a trifle rough, with a great deal of snarling and hissing. They could easily have been taken for real fights, except that neither cat was ever injured. But there was one unfriendly encounter between them – and it involved the kitten. Smudge had left him alone, and while she was away Pickle and Domino entered the outbuilding. Smudge returned unexpectedly and she was evidently very displeased. Domino made a run for it and escaped, but Pickle was trapped. This was the only time during the whole year when two of the farm cats had a serious quarrel. When it ended, and Pickle left, Smudge carried Lucky away from the outbuilding and into the barn and a new, safer – and incidentally probably more comfortable – nest. So it was the new home that Pickle took to visiting and very soon her help, for Pickle took

Lucky with his mother Smudge and grown-up sister Pickle

just as good care of Lucky as Smudge did, seemed to pacify Smudge and her attentions were accepted.

This may have looked like misplaced maternalism on the part of a cat that had lost its own kittens recently, but it was probably much more than that. Lucky was related to Pickle more closely than even Pickle's own kittens had been – not that Pickle could have known this. It was in her own interest to contribute to the survival of any kitten within the group. She did what she could – which was quite a lot – to make sure Lucky had a good start in life.

In her trips away from him, Smudge must have been catching food, and by the time Lucky was a fortnight old she may have been bringing some of it back for him. Lucky would not have been ready to start taking solid food until he was about a month old, but before that he would have been given food to examine. It was important to his education that he learned to recognise food as soon as possible.

A young kitten develops quickly, doubling its weight in the first week of its life. At five days, its shoulders, pelvic girdle and limbs have become strong enough for it to hold its legs beneath its body, rather than to the sides. It cannot walk yet, but it can grip the floor and it is more stable. After about twelve days, a kitten can sit up, supporting the front part of its body with its front legs, and it can use its hind legs to prop itself up. At about seventeen days, it will take its first steps. By then, its legs are strong enough for it to walk, but it takes a few days for a kitten to master the rather tricky balancing act that is needed as its weight is shifted from one foot to another. It falls over a great deal. At three weeks it can walk, sit with one forepaw raised and moving freely, and scratch its head with a hind paw. Its balance is less than perfect, even so, and it tumbles over frequently. By twenty-three days it is almost ready to explore the world. It can sit and wave first one front paw and then the other. It can raise itself on its hind legs and stand on its toes, but only briefly, because this is very tiring. It can climb a little, and it can and does squat to urinate and defecate. At these times it will scratch the ground with one forepaw and, if the surface is loose enough, make an honest attempt to bury its droppings. Most of the time, though, its mother continues to keep it clean. When it needs to urinate, it mews loudly and searches for a suitable place, but when its mother approaches it rolls on to its back for help.

Kittens start 'milk treading' within a few days, working away at the belly of the mother, stretching first one foreleg and then the other, claws extended until they meet the mother's skin, when they are retracted and the leg bent. The process stimulates the milk flow and is accompanied by loud purring. Sometimes the mother will perform similar movements and, of course, they are familiar to every cat owner. They are performed by both males and females, often throughout their lives. What some cat owners may not have noticed, however, is that the movements involved in milk treading are virtually identical to those used in claw sharpening. The purring is omitted, it is true, and the movements are smaller, but the reduction in their scale may be caused by the resistance a rough surface offers to the claws.

Is claw sharpening developed from milk treading? No one knows, but the purpose of milk treading in young kittens is fairly straightforward. When the kitten is very small, its mother's milk flows so plentifully that the kitten merely sucks and swallows, obtaining milk as fast as it can manage. As it grows, and both its requirement and its ability to suck and swallow increase, the flow, which remains more or less constant, seems to have been reduced simply because the kitten can take it much faster. By kneading its mother's abdomen it increases the rate of flow. It is quite literally milking its mother. Calves will butt their mothers – often very hard – for the same reason. Like all young animals, a kitten has a limited range of demands, each of which is extremely urgent, and when these demands are satisfied the kitten is pacified and feels contented. If, as seems likely, the growing kitten comes to associate its milk treading movements with the contentment they produce later, this association of feelings may remain with it long after it has been weaned, and milk treading will produce contentment even though it produces no milk. Claw sharpening – in fact the removal of the outer layer from the claws – has a quite different purpose, but the similarity of the movements involved may lead to some kind of identification between the two activities, so that while it is not as satisfying as milk treading, claw sharpening also brings contentment. Cats seem to do it when some minor inconvenience has disturbed them, and they seem to derive comfort from it.

Self-defensive behaviour also begins early. A very young kitten, up to two weeks old, will approach anything that seems interesting. Until then its world has consisted entirely of its mother, and its approaches to her are rewarded. After that age, though, kittens are more cautious. An object that moves rapidly will make them retreat and arch their backs. They may hiss and spit – and it is not unknown for a kitten to hiss at a strange smell when it is only two days old.

Small kittens are restless, and their movements mean one of only two things. Either they are hungry and are looking for their mother, or they wish to rest. Lucky had no litter-mates with which to sleep, but had he had brothers and sisters of his own age, all of them would have preferred to sleep in a heap, their heads free and their chins resting on their neighbours. It is quite a complicated business for a litter of kittens to work out a configuration that will allow each member of the heap to rest its chin in the most comfortable position, and the operation can take some time, during which the kittens wriggle and writhe until suddenly the magic position is found. At that moment they all fall asleep simultaneously. Of course, they continue to move in their sleep, and that is how fights often start. By the time they are three weeks old, sleeping kittens are liable to wake up at the bottom of the heap. When that happens, they wriggle out backwards and work their way back to the top. This means climbing over the top of other kittens, and the kitten whose rest has been disturbed has little respect for the undisturbed peace of its fellows. The climber will slam the other kitten with alternate movements of its forelegs, claws extended, and finally it will fall on to it with its mouth open. As it closes its mouth it bites the other kitten. At that age the bites do not hurt and the claws do not really scratch. Its arrival at the top may well mean that another kitten is now at the bottom, and the process may continue until it degenerates into a free-for-all.

The kittens are flexing their muscles and it is essential to their proper development that they should do so. The young body must be stretched, exercised, used, and the control of it learned. This is not taught – 'formal education' begins later – but it must be learned, and some of the habits that begin during this early phase remain with the cat for the rest of its life.

The kitten soon learns to jump vertically into the air. It is not very difficult, after all, once it has mastered the skill of moving all four legs in unison. As soon as it can jump, it can jump *somewhere*. Somewhere is usually on to a litter-mate, and the litter-mate resists. So a play-fight begins. If the victim of the attack should roll over, so that the attacker's hind legs are lifted from the ground, the attacker may begin to kick its opponent. There is not much difference between the tactics of attack and defence. It largely depends which combatant is on top, but if the defending kitten rolls on to its back, it finds it has freed all its four paws for use simultaneously, so it bites, swipes with its forepaws, and kicks all together. Rolling on to the back has two distinct purposes, but because the movement is identical in both cases, one purpose often turns into the other. The kitten was rolled on to its back first by its mother, to be cleaned. Then it signalled its need to urinate by rolling on its back, so that its mother would help it. Gradually it comes to lie on its back as a means of encouraging pleasant encounters with its litter-mates. If it lives with humans, it may roll on to its back to encourage such encounters with them. Not every adult cat does it, but most cat owners are familiar with the pet that likes 'to have its tummy tickled'. (Dogs also do this, and for the same reason.) It is infantile behaviour – but does it imply that so far as the cat is concerned, its human companion is a litter-mate?

At the same time, though, lying on its back is the best position for defence, and when it becomes bored with having its tummy tickled it is the most natural thing in the world for the kitten to turn the encounter into a fight.

It is at this very young age, too, that the kitten learns to greet friendly cats and kittens. When they awake in their heap, kittens often groom one another, and as they learn to walk so they learn to greet one another. Attacks usually come from the rear, and so a kitten that is met head-on is unlikely to behave aggressively. The two kittens touch noses and then pass one another, briefly touching flanks as they do so, with their tails held high but each tail curved slightly towards the other animal. Before long this greeting develops into a more serious one, and the kittens rub their cheeks against each other's flanks, in behaviour that is typical of all domestic cats. The cheek rubbing is a form of scent-marking, and by repeating it frequently all the members of a litter come to possess a common smell that becomes an important means of recognition when they begin to wander from the nest. They scent-mark inanimate objects in the same way, and this is probably a form of 'mapping' that will help them find their way.

By the first week in July, Lucky could be heard – by Peter – moving about inside the nest. He was getting ready to begin serious exploration of the world outside. Both Pickle and Smudge were behaving aggressively towards any cat that intruded, and while Pickle was spending less time with Lucky than she did during his first fortnight, she continued to defend him, chasing away cats whenever Smudge was absent.

Until they are about a month old, all kittens are very much alike in their behaviour, but after that individual differences begin to appear and the kittens start to acquire 'personalities'. Lucky had no kittens to play with, and so he played with the adults. It was less satisfactory than playing with other kittens, though he was not to know that. The most usual playmate was Smudge, but while she might respond by joining in the play, she might equally well place a paw on Lucky's back and draw him to her, whereupon he would forget the game and start suckling. There were advantages, too, in his reliance on adult companions, for unlike a kitten whose world consists mainly of other kittens, he was able to watch the adults and he learned by imitation. Peter watched him acting out little parodies of Smudge and Pickle as they stalked or chased prey. By the third week in July there was no evidence that Lucky had begun to eat solid food, but he was sniffing, pawing and chewing at objects he encountered. Should he come across something that tasted good, he was capable of eating it.

His behaviour was mainly exploratory, for he had reached the age at which a kitten first exhibits the curiosity for which cats are justly famed. Curiosity is of the greatest importance. The environment, and everything in the environment, must be examined carefully. Secure sleeping places must be found and their positions noted. Anything that might be a source of food must be tested, but with caution. Caution is also important, for despite its ability to rescue itself from scrapes, curiosity carried too far really can kill the cat. It is very unlikely that when Pickle found the slow-worm she knew anything about snakes, but she was cautious and her caution was wise. The reptile was quite harmless, as it happened, but had it been an adder – and in the part of the world in which the domestic cat first appeared it could have been a cobra – a careless attempt to seize it could have proved disastrous.

Lucky ventures alone into the outside world

Much of Lucky's play resembled fighting. He was learning to run, and to jump away from his opponent as well as towards him or her. He was lying in ambush to mount his attacks and, somewhat precociously, he was beginning to stalk. As he grew, his technique improved in well marked stages. At first he would merely romp, jumping playfully at any cat that moved. Soon, though, he learned that his bites and scratches could inflict pain, and it became necessary for his attacks to be aborted at the last moment. He would stalk, chase or run, but stop just short of the final leap. Until his co-ordination improved, the abrupt halt made him fall. He began stalking much more and the old rushing attacks became less frequent. He started to use cover, and he learned to bend his knees so he could crawl forward slowly, his belly almost touching the ground. Sometimes he would stop, with his hind legs drawn forward, and his whole body would quiver from side to side, while his tail made a slow waving movement. This stance is a familiar one, used by all cats when they are hunting. Probably the movement of the tail helps maintain the animal's balance, allowing for quick changes in balance that may be needed as the attack begins. Probably the movement of the body and head help the cat to estimate the distance and precise location of the target. It has been suggested, though, that they begin not as a means of improving the efficiency of

Lucky listening to every sound . . .

the attack, but as a nervous response to the inhibition of the kittenish rush, as a kind of frustration that is later turned to good use. Be that as it may, Lucky's crouch would end either with a jump or with more stalking. He was now nearly six weeks old.

He played with both Pickle and Smudge and although the time he spent with each of them was more or less equal, they behaved quite differently towards him. Peter noted this difference:

Quantitatively Pickle and Smudge interact with the kitten just as often. Qualitatively there are differences. Pickle indulges in much more rough and tumble play than does Smudge and some of the sessions look positively violent, but the kitten keeps going back for more. During licking sessions Pickle sometimes bites and holds the kitten for short periods. This makes it lie still for a few seconds while it is licked. Smudge uses a far more effective method of trapping it in her forepaws or lying on top of it. In view of the importance of allogrooming in the adults it is interesting to see that the kitten is already showing this behaviour. . . . Its development is doubtless influenced by the lack of other kittens to play with.

Domino tended to avoid Lucky. She would not play and, as often

. . . and reacting to the slightest movement

as not, she would jump away and growl whenever he approached her. She was pregnant again, and towards the end of July she was looking pregnant, although Tom still found her attractive enough to bite her neck and mount her on the 16th. She had become sexually receptive soon after the death of her kittens, and Tom lost no time in attempting to father a new generation by her. Although neither of them could know it, of course, they could not afford to waste time. The year was advancing, and the kittens that are most likely to succeed in life are those that are able to hunt for themselves – at least partially – by the time the winter snows blanket scents and the small mammals retreat deep into their burrows. During the winter an unweaned and helpless kitten is a severe liability.

Domino's attitude to Lucky may have been caused by confusion rather than by her pregnancy. It is possible that she did not recognise the kitten as a member of the family, at least not when he was playing. Peter suspected this, because he observed that while her reaction ranged from mild hostility to friendly licking, she seemed much better disposed towards Lucky when he was sleeping with Smudge. 'Maybe then she recognises it as a member of the group?' he speculated.

Tom was not very impressed by Lucky, either. 'Tom initially growled at the kitten every time it came near and would sometimes

129

move away. The kitten sometimes responded by defensive crouching and hissing. At the end of the week Tom was much more friendly and licked the kitten on three occasions – Maurice and I were trying to get some shots of unfriendly interactions at the time!' Biologically, Tom's task was done. The kitten was alive and well, and its upbringing was strictly the business of the females. He never hunted for Lucky or brought him food, and it was rare for him to play with the kitten. In fact, as Lucky grew bigger and stronger, Tom moved closer and closer to the edge of his range and the other cats saw less of him. It was as though Lucky was displacing him in the real world, just as he was displacing him as the centre of attention for the human observers.

Peter had not lost interest in Tom, by any means, and he was puzzled by the long absences. One day he managed to track Tom for some distance, but then lost him as the cat disappeared over the top of a hill which blocked the radio signal. However, his secret assignations were eventually disclosed and Maurice even managed to film one of them, a shot of Tom walking into a barn on the neighbouring farm, tail held high in greeting as he approached the lone female called Broken Ear. It was then that Peter discovered that Broken Ear had given birth to two kittens which, by their fur markings, Tom had almost certainly fathered. Whatever might happen to Lucky, Tom wasn't taking any chances with his precious genes. The wider his circle of female acquaintances, the more certain his personal influence on future generations.

Lucky was roaming all over the barn. From the age of six weeks a kitten will begin to climb suitable objects. It has little difficulty going up, but descending presents real problems, and faced with a difficult descent a kitten is quite likely to go further up instead, because it is easier. Although they climb so easily, they are often scared of heights, but eventually each individual devises its own means of descending. On a tree with waving branches a kitten may jump haphazardly from branch to branch, sometimes landing clumsily, sometimes actually falling. With a little practice, though, it learns to co-ordinate its leaps with the movements of the tree, so it can descend gracefully in this way as far as the lowest branch. Some kittens learn very quickly to do this, showing a distinct improvement after only one attempt. For the last stretch of the trunk, the kitten may either try a head-first descent or choose to descend head uppermost, which is less elegant but a lot safer.

Although cats do not know it at first, a fall will always land them on their feet. Indeed, the ability of a cat to land on its feet from a fall is legendary. How cats achieved it puzzled people for centuries, and it was not until the invention of the ciné-camera that the puzzle was solved. A very famous film sequence, made in 1894 by a French physiologist named Marey, shows a cat falling. The action can be examined frame by frame. The film revealed that the secret lies in the ability of the cat to twist its body about its own longitudinal axis. It cannot right itself by twisting at right-angles to that axis. For physical reasons it is virtually impossible for a cat to correct 'in flight' a strongly head-up or head-down attitude. For that it must wait until it touches the ground with the two paws that are in the lower position, but then it falls fairly gently and gracefully on to all four paws.

If you imagine a cat in a more or less prone position, with all four feet pointing upward, it will begin to correct its position by extending its hind legs fully and drawing its front legs as close as it can to its body. This makes it possible for the cat to twist the front part of its body over until its front legs are below it. Its hind legs now point to one side and slightly upward. It stretches its hind legs directly behind its body and twists the rear part of its body around so its spine is untwisted. It brings its hind legs forward until they point downward, extends its front legs so they, too, point downward, and finally it arches its back. It is now in the correct position for landing.

The truly remarkable aspect of the feat is the speed with which it is performed – and that is why it was so long before anyone discovered how it was done. A cat can fall from a height of three feet with its feet pointing upward and still land on all four feet. This means it corrects its position in about one-eighth of a second.

Once it had been discovered how the cat lands on its feet, the next step was to discover which senses it uses. At one time it was thought the tail was used to help the cat balance itself, but that idea had to be abandoned when it was found that a Manx cat can land on its feet as well as any other cat. In the end, it was agreed that the cat uses the semi-circular canals associated with the ears, and its eyes, but that the eyes are the more important.

Cats can fall from considerable heights without being injured. The terminal velocity of a cat falling freely in still air – the speed beyond which it will not accelerate further – is about forty miles an hour, and at that speed it can land safely provided the surface is even.

Lucky was well equipped with keen senses and reflexes. By the end of July he was ready to learn the principles of hunting. Smudge would bring prey to Lucky, announcing her arrival with a call that brought him hurrying to her. At this stage the prey would still be alive, and she would drop it in front of him. At first he would sniff and, if the prey looked as though it might try to defend itself, he would back off from it. Then he might lick the animal, or bite it, and if it moved away from him he might pursue it. Until he learned to kill, Smudge would watch him for a time and then catch the prey again herself, kill it, and drop it in front of him once more. This time he might try to eat it, but if he did not, Smudge would begin to eat it herself and would permit him to share in the meal if he wished.

He was learning to concentrate for longer periods, and would chase any small moving object. Butterflies, leaves, and any other passing fancy would set him leaping and, when his mouth could not reach the quarry, he would swipe at it with his forepaws. He soon became bored, though, and it was some time before he was able to see his chase through until either he had caught the object or his hunt was clearly hopeless. Smudge's lessons were very important, because young kittens must learn to kill. The urge to chase is innate, and a cat will catch any small animal it can, but it must be taught how to kill it. It takes practice for a kitten to learn a method that works, and when it succeeds its behaviour becomes stereotyped and the cat will not vary it. The prey must be manipulated until its head is held forward and its body is fixed securely between the cat's forepaws. Then the bite to the neck can be delivered. By the time he had learned this, Lucky's games

Smudge was a patient and attentive mother to the growing kitten

with moving 'toys' and the lessons he had received from Smudge had taught him not only how to control his own body and organise himself for the hunt, but how particular prey behaves. He began to recognise prey species by the way they moved.

Smudge was a good teacher. Not all mother cats are so skilful, and some kittens never learn to hunt and kill really efficiently. In nature, of course, such families would be eliminated, but domestication has encouraged the survival of cats that are unable to hunt well, or even to hunt at all. One study of cat behaviour noted a mother cat whose lessons were brief and who always ate her prey herself. Her kittens would cower in a corner should a mouse enter their nest. Probably the mother's inadequacy as a teacher resulted from the way she herself was taught as a kitten. It is while they are learning to hunt that kittens also learn to behave cautiously when the prey is unfamiliar and possibly dangerous.

Early in August Lucky's hunting and killing lessons from Smudge were going well, but sometimes Pickle interfered. Peter watched one day while Smudge brought in a live mouse and gave it to Lucky. The kitten played with it using neck bites, trapping it with one or both paws, and throwing it into the air. He did not get to kill it, though, because Pickle dashed in and stole it from him before he had a chance to do so. She carried it away, growling loudly, and ate it. Clearly there is more to a kitten's education than just hunting! Smudge made no attempt to recover the mouse, but went outside, caught another, and gave that to Lucky – although eventually she ate it herself. Peter had no doubt that Lucky would grow up a competent hunter, though so far he had seen the kitten catch and kill nothing larger than a moth. His diet was not exclusively carnivorous. A couple of days after the incident with the mice, Smudge arrived in the barn carrying a lump of Edith Tibbles' chocolate cake, taken from the dustbin, and Lucky ate some of it. He was drinking milk regularly and there were no fears that he would starve.

He was an 'only child'. Lacking other kittens his behaviour was not entirely typical. His only playmates were adults, a situation that is unusual among domestic cats, where most litters consist of at least two kittens. The adults were not always willing to play with him, but he pestered them constantly and so dominated life in the barn. It may well have been his pestering of Tom, who hardly ever played with him, that made Tom seek a more congenial life elsewhere. Lucky could not be ignored, for his importuning took the form of flying leaps at the adults. Sometimes he would come off worst and be driven away gruffly, but

Lucky, 'an only child', learned play from his grown-up sisters

there were no prolonged sulks. Smudge was more tolerant of him than either of his older sisters, and as time went on Shadow, too, became a playmate of a unique kind. Peter observed that 'the playful interactions with Shadow are all chases – he is about the only thing which ever runs away from the kitten.' Smudge took more notice of Lucky as Pickle's special relationship with him drew to its close, and as Domino became less hostile she and Pickle began to behave towards him in much the same way.

Domino was now heavily pregnant again and had been sleeping in the nest she had made for herself in the barn. She was preoccupied with her own condition for much of the time, and at ten o'clock in the evening of a Monday late in August her litter started to be born. It was another episode that ended in tragedy. Peter described it, and his own attempt to assist, as dispassionately as he could.

At about 2140 Domino was still wandering about showing contractions and mewing, so I put her in a handy nest. Ten minutes later the contractions got stronger and she began to mew more loudly. Smudge was attracted and sniffed at Domino who began to stagger off towards the shed across the road. In the concrete yard the contractions became too strong for walking and the first kitten was born. Domino licked it but began walking about, dragging it along by the umbilicus. She seemed unable to reach it easily and walked round in circles trying to lick it. Smudge was attracted by its distress calls but licked the kitten only briefly. To save the kitten from injury, I cut the umbilicus as near to Domino as I could. Domino licked it a little more then carried it over to the shed. By 6 o'clock the following morning she had had two more kittens. One of these was suckling strongly, the other was dead. The eldest kitten was barely alive and showed little response to an hour's warming and attempts to make it suckle. At 0900, however, both kittens were suckling strongly. Domino groomed the kittens and allowed them to suckle, but the younger of the two died at 0100 on the 23rd and the other was dead at 0900. After feeding, Domino went back to the shed only once and showed minimal interest in the kittens. Since then she has behaved as if nothing had happened.

Lucky continued to be the only survivor, and he was growing up fast. After the death of her kittens Domino spent more time with him, was more affectionate towards him, and he usually slept with Domino and Smudge, all together. He was becoming less playful. It was not

Lucky's keen senses equipped him to survive

only that he played less with the other cats; he also spent less time in the hunting and chasing games he made up for himself. He began to go with Smudge on her hunting expeditions to the chicken house. He did not stay for complete sessions, and so far as Peter could tell he did not catch and eat prey himself. The day when he did so, though, could not be far off, for the way he dealt with the prey Smudge gave him was now more purposeful. He manipulated it, pounced on it, and delivered neck bites. Smudge gave him mice that were still alive, but she made sure that the rats she caught were dead when she presented them to him. A rat would have been too dangerous an opponent for him.

His luck continued to hold, for he caught flu and recovered from it. It kept him confined to the barn for a time and he obviously felt unwell, but the illness passed. Domino, too, spent much of her time in the barn, and she and Lucky became close companions, playing and sleeping together more often than either of them did with any other cat. Lucky had learned to catch and eat crane flies, and he developed a taste for especially fat, large spiders.

By mid-September, although he continued to suckle now and then – long after a house cat would have been weaned – he was increasing his range, and his social position among the cats was becoming more like that of an adult. He was passing into a new phase of his life. He had been an entirely social animal up to now, totally dependent on the company of the other cats, but as he became better able to fend for himself, so he became more solitary. It was not a sudden change. In fact it was many months before it was to be complete, but the beginning of it could be perceived. One day, perhaps sooner perhaps later, Lucky might well leave home to become for a time a solitary hunter. He was still mostly kitten – the change was not much more than a hint of what was to come. He was very friendly and sometimes still playful, but one aspect of the change could be seen in his attitude, some of the time, to Smudge. When she entered the barn, for example, he would greet her – like an adult – before she could start grooming him. By the end of September Smudge had almost succeeded in weaning him. He did not want to be weaned and kept trying to suckle, but she resisted his attempts, sometimes by walking away from him when he tried it, and finally by behaving aggressively towards him.

The battery of Smudge's radio collar failed early in October. The radio was still attached to her and transmitted intermittently, but it was no longer reliable. She and Lucky disappeared and were gone for the whole of one night and all the following day. Peter found them at last, beyond the limits of Smudge's known range. Peter supposed that Smudge had taken the kitten there, but it was just possible that Lucky had gone there by himself and Smudge had gone looking for him.

By late November Lucky was treating Smudge in much the same way as he treated the other farm cats. He spent a good deal of his time with Shadow. He seemed to enjoy his company, but probably that was because Shadow was frightened of him and ran at his approach, so he could enjoy chasing him.

As winter began, the midsummer kitten was large enough and well enough trained to begin to fend for himself.

The winter

IT WAS early in November that Smudge was injured on the road. Her right front leg and lower jaw were so damaged that she could not hunt for herself, let alone provide for her dependent offspring. For Lucky it was to be his first practical lesson in survival. Peter was worried about him and he had cause to be, for winter was approaching. Confined to her 'sickroom' in Peter's care, Smudge was unable to watch over Lucky and he needed supervision. No one had ever seen him catch for himself anything more substantial than a spider and it seemed very unlikely that he would be able to feed himself. But to everyone's surprise, Lucky remained in rude health. In fact he put on weight. It was a mystery how he had so suddenly become such a successful hunter.

It was Edith who solved it. Every day she fed the chickens. One day, having emptied the bucket in their trough, she paused to watch them eating. One of the chickens was not a chicken at all! Lucky had decided that he was a match for any bunch of clucking hens and had discovered that their rations included tasty morsels of meat. He took to shouldering the chickens aside and joining them at their table, seizing the meat for himself and leaving the corn for them.

The incident delighted Peter and Maurice and it had to be filmed. This was simple enough. All Maurice had to do was set up the camera and wait for feeding time. A few minutes after the chickens had been fed, in trotted Lucky for his breakfast: his cunning antics were recorded for posterity.

This was all very well, but despite his rapidly increasing size and his growing independence, Lucky was still very much a kitten. His education was not complete. In the height of the summer, when the young of many prey species are available, he might have stood a fair chance on his own. As it was, his inexperience would tell against him. It takes little skill, after all, to catch a fledgling that falls from a nest or a young mouse that wanders and does not know how to return to safety.

Lucky raiding the chicken trough; Shadow, the shy outsider, looking on

By early winter the previous season's young would be well advanced and they would be learning, faster than Lucky was learning. He had been born fairly late in the year, and their skill at evading capture might well exceed his skill as a hunter.

Winter is the great leveller among wild animals. With food in short supply, all resources are strained. It is the time when the weak, the sick, and the unskilled are most likely to starve, leaving behind a nucleus of the toughest individuals to breed in the coming year. It is not merely that the dying down of plants leaves the herbivores short of food, although that is serious enough. The insects survive the winter mainly as eggs or hibernating larvae, and there are few adults. This forces some insectivorous species to turn to plant foods, but many more migrate – especially the birds. Birds are not really important as sources of food for cats, but they do feed other predators and these must now turn to small mammals. In one way or another, the hunting cat must search more diligently in the face of increased competition. Among the predators, success now was to the swiftest, the most patient, the keenest of eye and ear, and the most experienced. The world in winter is no place for a kitten on its own.

If you imagine all life in the countryside arranged like a pyramid whose shape remains constant no matter what happens, the significance of winter will become apparent. The plants must form the base of the pyramid, for directly or indirectly all animals feed on plants. Above the plants, then, come the herbivorous animals and, above them, the carnivores. Any change in the quantity of plant food will alter the whole of the rest of the pyramid – for if herbivores starve there will be less food for carnivores, and if herbivores thrive and multiply the carnivores will thrive and multiply also.

The interesting corollary of this is that only rarely, and under special circumstances, is it predation that regulates the population of prey species. In the real world their numbers are regulated by the availability of their own food and such other essential resources as nesting sites. It is the size of the prey population that regulates the size of the population of predators that in turn feed upon them. Far from the number of cats in a neighbourhood governing the number of mice, therefore, it is the number of mice that determine the number of cats and the other hunters of mice.

Are farmers fools, then, when they 'employ' cats to keep down the numbers of mice and rats? Because farmers supply virtually inexhaustible amounts of plant food, there can be many mice – and cats – but their barns also tend to exclude other predators. The natural population controls are removed, and so cats really do exert an influence on prey population. The farmer who keeps cats is a prudent man.

The Devon cats controlled vermin very efficiently. Maurice noticed the effect, especially after the project had ended. Once the cats had gone, the mice and rats began to multiply again.

Lucky, though, was at risk; Edith was now wise to his antics in the chicken run. On 12th November Peter recorded that Lucky had been scaling the walls of the shed in which Smudge was convalescing: 'The attraction is probably her food rather than her company, although he has been suckling again. I have put up some netting which so far has kept him out. As far as I know he has not been hunting.' By 16th November Smudge was walking on her injured paw, but the kitten was becoming thinner. Peter's anxiety was evident in the note he made: '. . . he was hunting in the chicken house, although he did not catch anything – success for the adults is rare anyway. There are several mice in the sheds so he should get one sooner or later. Lucky's range is

Lucky became a hunter 'at the eleventh hour'

restricted at the moment to the barns and the orchard.'

With Smudge temporarily out of the way, Lucky was spending more time with Domino. The two cats slept together often and Domino sometimes groomed the kitten without prompting. Peter found this a little puzzling; perhaps Lucky needed a wash!

One day late in November Maurice and Peter had been following Lucky to film whatever he might get up to by himself. They saw him on a low embankment, evidently stalking something. As they watched, he prepared himself, then leapt forward in a kind of lunging pounce. It lacked confidence, it was clumsy, but it succeeded, and Lucky had a mouse. He might have caught one before, but this was the first time anyone had seen him catch prey, and the excitement of Maurice and Peter who watched and filmed the event was tinged with more than a little relief. Lucky had become a hunter at the eleventh hour. Now that he had accomplished the skill, he would improve day by day.

The first frosts saw Tom ranging more widely across the country-side and making frequent visits to the village. The cats whose lives continued to centre on the barn saw little of him, and when he did return his mood was sometimes tetchy. One day Peter saw Tom and Pickle meet by the gate that opened into the field by the barn. Tom had travelled a long way, to the edge of his range beyond the village, followed by Peter, and he was on his way back. Pickle was hunting by the gate when Tom arrived, but they were not alone.

A very large ginger cat, probably the pale one from the manor, trotted quickly north from about a hundred yards away. Pickle stared hard at it but Tom was apparently unaware of its approach until it was about twenty yards away. Tom then looked briefly at it and crept away with his ears back until I was between him and it. Ginger walked through the barn field gate, passing Pickle on the way, and hunted south along the edge of the field. Pickle and Tom followed at a distance of about ten yards. Ginger turned and walked north. Pickle crouched in the hedge ready to pounce but allowed Ginger to walk past. As he did so he looked at her and increased his pace. Tom seemed to gain confidence and trotted after Ginger as he walked under the gate. Tom did not pursue him past the gate. Ginger disappeared northward up the road. Tom and Pickle returned to the barn where Tom worked off some of his aggression on Domino and Lucky – snarling fiercely when they approached.

This kind of behaviour was unusual, though. Most of the time Tom was friendly enough. Once Peter saw him sleeping with Pickle. Domino went and joined them, leaving Lucky to sleep by himself a few yards away. When Tom left again, Domino stayed with Pickle and later, when she saw Pickle sleeping by herself, she joined her again. After that episode Domino and Lucky spent less time together. Their 'special relationship' seemed to have ended.

Although the female cats had been seen to share food, and although they brought food for the kitten when he was young, Tom never helped any other cat. He never provided food for others and he played no part in the upbringing of Lucky.

Tom's disinterest – which never amounted to hostility, despite his occasional irritation with Lucky – may have seemed selfish, but was it? Biologically, what mattered was the survival of Lucky, and the survival of Tom himself, who would live to sire more kittens. Could he have done more for Lucky than merely father him? Had he stayed to help with the rearing, what could he have contributed? His help in grooming and supervising the kitten would have been superfluous, since Smudge was quite capable of supplying as much care of this kind as was needed and, as we have seen, she had the help of Pickle and Domino. Tom could have brought food to the nest, but in order to do that he would have had to make a choice. If he obtained the food from the places he knew and the sources he exploited with confidence, he would have had to carry this food over long distances, because most of his range lay beyond the ranges of the females. There would have been a serious risk of losses, not least from highway robbery by rival cats. What is more, he would have expended additional energy moving to and fro, and so his own demand for food would have increased. All in all, it is far from impossible that he would have brought home rather little food and would have gone hungry himself. His alternative, then, would have been to work a range much closer to the barn – inside the ranges of the females in fact. Perhaps he could have provided useful amounts of food in this way, but since he could not have worked both the close range and his own more distant range, the close range would also have had to feed him as well. The net result is that the group would have been worse off for his help. Tom did not plan any of this, of course. He had no concepts of 'selfishness' or 'unselfishness'. Tom merely did what he did, and although he may have appeared as a bad father, a more careful consideration of his behaviour reveals the fact that its effects were beneficial to the group as a whole.

Tom had other interests in any case. If we allow that his main business was to father new generations, the reason for his visits to the

The complete family in the barn

Peter and David tracking the cats over the winter countryside

barn appear in a different light. He had to call on all the females of his acquaintance from time to time to discover whether they were sexually receptive. His visits would require the formalities that attend encounters among friendly cats, and so he would greet his females and display what any human might interpret as affection for them. He might sleep with them, as he did with Pickle, because, after all, he had to sleep somewhere and cats prefer to sleep in a heap if they can. They keep warmer that way and they are less vulnerable to predators, because in a heap there is a greater probability that one animal will wake at the approach of danger. Perhaps cat heaps work because within them there is always one cat that cannot get to sleep at all!

As time went on and Lucky grew, Tom roamed further afield. Lucky was still a young cat when the Devon project ended, and not sexually mature, but in time it seemed likely that he would take over Tom's females and would keep Tom away from them.

What Peter, David and Maurice saw was the behaviour of cats that live semi-independently of humans. House cats behave differently – but there are many similarities. Many viewers sent in descriptions of co-operation in the rearing of young between females, but only a very few had seen co-operation that involved males. In one case this involved a pair of cats that had themselves been raised together and the co-operation amounted to little more than the male playing with the kittens. Even Tom did that occasionally! A more curious tale concerns a male who broke a leg at just about the age when he may have been maturing sexually. He lived in the same household as his mother, and when she produced a litter he was seen grooming them and later he appeared to be trying to suckle them. Here, though, it seems most likely that his behaviour was that of an 'aunt', a sexually immature individual who assists by minding the young. Another viewer described two males, one of which had been neutered. When the unneutered animal was injured, the neutered cat brought food for it and defended it, but later, when the injured cat had recovered but the other one was hurt, the compliment was not returned. In yet another story, a full brother and sister produced a litter and shared their upbringing. In this case – which might well have become Lucky's case had he remained to mate with his sisters – the genes inherited by the kittens would be merely a reshuffling of those possessed by both parents, which themselves were shared. Each generation did not reduce the degree of relationship among individuals, as it would if genes were introduced from outside.

Lucky was approaching the age at which he might have acted as an 'aunt' to younger members of the group – were there any such younger members. By December he treated Smudge in much the same way as he treated the other cats; he played less and hunted more. He was an immature male, an adolescent. Peter still provided food from time to time, usually by leaving it where he knew it would be found. It was always accepted, because no source of food could be overlooked.

The winter landscape was starting to look bleak, but appearances can be deceptive in some ways. It is during late autumn and early winter, for example, that the population of wood mice reaches its peak. Oddly enough it is at its lowest in midsummer. What seems to happen is that during the latter part of the winter, when the weather is often at its most severe and food stocks are at their lowest, mortality among the mice is high. Even so, it is rare to find an undernourished mouse – probably because the sick and the weak die quickly. In spring there is an outward migration. Some mice live in one place all their adult lives, but there are others that never settle for long in one place. During winter these 'transients' find shelter, but as the weather improves they resume their travels, joined by some mice born the previous year that are setting off to find new places to settle. The population falls still more as the breeding season begins and males become aggressive and drive out rivals. The birth of young should cause the population to increase, but infant mortality is so high in the early part of the year that births can do no more than hold the population size constant. During the late summer and early autumn the situation changes as increasing numbers of adults die from old age. More young are born and survive, and the population starts to increase once more.

The fact that the habitat might actually support more mice does not mean that the cats will be able to catch them. As winter sets in the mice become less active, and if snow falls they become almost impossible to find. Snow makes life easier for many very small mammals. It doesn't lie on the ground but on the top of the herbs that stand an inch or two above the ground, and mice and voles can move easily beneath it. The cat, sitting above, will not smell them because the snow masks their odour, and it may not even hear them, for the snow muffles the sound.

A few mammals hibernate. The number of true hibernators is much smaller than many people suppose, for hibernation is a very difficult physiological feat. It requires an animal to override that part of the brain which acts as a 'thermostat' and triggers the activity – internal and external – by which body temperature is held steady. Temperature drops, heart beat slows, respiration very nearly ceases and, most difficult of all perhaps, toxic waste products from the body are not excreted but stored where they can do no harm. Hedgehogs and dormice hibernate, but wood mice, voles and squirrels do not – although they do spend much of the winter sleeping.

By allowing its body temperature to drop to within a few degrees of the air temperature, an animal greatly reduces the amount of food it needs to survive. Most of the food eaten by mammals – including humans – is used as 'fuel' to maintain body temperature and respiration. Remove that need, or most of it, and you remove the need for most of the food. As the winter of the temperate regions tightens its grip, hibernation is obviously a convenient strategy for those animals that can manage it.

If we take a kind of inventory, then, we find that the migrant birds have gone. The resident birds are difficult to catch because they tend to feed in flocks during the winter. The flocks are often of mixed species and they may be very large indeed. Each bill that feeds has, just above it, two eyes that watch. The instant a bird spies a predator it utters a warning that sends all the birds into the trees. There are no fledglings, of course, and there are no birds that have not had several months' experience at keeping out of trouble. The cat may watch, but only rarely will a bird fall victim to it. The small rodents are still in residence, and apart from the hibernating dormice they move about occasionally, but only occasionally. For much of the time they are in their nests, and inaccessible. The cat may sit and wait, but it may be a long time before he catches a glimpse of a mouse.

At the same time the predators remain active, and hungry. The buzzard sits in its elm, or hangs on the daytime air, and the owl watches and glides by night. The fox prowls by the hedges and across the fields in search of a meal. Even the weasels and stoats, in their white winter coats, continue to hunt. It is against this competition that the cat must hunt, and also keep itself warm.

The cat is not immune from cold. It can suffer from exposure just as a human can, although it is less likely to make the elementary mistakes that lead to hypothermia. It has no special protection, though, and it cannot even retreat into relative inactivity like the species that comprise its prey. With luck (perhaps design) it will have no dependent young to care for during the winter months, but that is the only concession it is allowed.

Cats that live in the wild grow thicker coats for the winter. This thickening of the fur occurs in many species that inhabit the high latitudes, but when they live with humans, whose houses are kept at summer temperatures throughout the year, it is less marked. Lucky's fur grew thicker and longer as winter drew near. The change in his appearance was very evident. It was the only preparation he did or could make.

It was a hard winter. A cold spell in November was followed by mild weather that lasted until Christmas, but on New Year's Eve heavy snow sealed the Devon landscape. The road from the farm to the village was blocked several times. The wind blew sharp and bitter round the barn.

Peter was forced to stay in his hut much of the time, venturing from the farm only when his supplies ran low. He behaved much as all the other animals behaved. The cats remained in the barn, snuggled together for warmth among the bales of hay. When hunger forced him to hunt, Lucky worked mainly in the churchyard, and he drank from the pond behind Peter's hut.

It was not only hunting that was difficult in the snow. For Lucky it was difficult even to travel. Each step was a struggle as he sank deep into the treacherous ground and had to fight to free himself for another step. Now and then he would climb a tree, which at least was solid, and rummage into holes in the bark in the hope of finding a sheltering bird, mouse or squirrel. Sometimes he succeeded, but more often he failed.

Lucky on the prowl; he soon expanded his own home range

Smudge continued to share her food with him, which helped, for she was still the more skilled hunter, and Peter provided the occasional accident victim in the form of a rabbit or squirrel collected from the road. In the tradition of the true opportunist, Lucky learned to steal where he could. Maurice once left the window of his car open and one of Edith's famous fruit cakes on a seat. He returned to find Lucky inside the car and most of the cake reduced to crumbs.

But for most of the winter Lucky remained independent. He demonstrated that he was an adaptable creature – and certainly not shy of exploiting the humans who shared part of his home range. Foxes will feed in the same way, far into cities where there are few hedges and even fewer fields for them to patrol for their prey. It does not mean the visiting animals are 'tame', of course, although their hunger may make them more easily approachable. In their world, humans are just another animal species that occupies a habitat they can exploit. They can steal food from humans, that is all. In this winter exploitation we can perhaps see a prototype of a more elaborate and full-time exploitation that extends to the shelter of human dwellings and capitalisation on their hospitality. They are the same animals, the cats that purr by the fireside and those that wander alone on the hillsides and hide from man. The one shades imperceptibly into the other.

Although the winter was harsh, it was short. The south-western peninsula of England, washed on two sides by the sea, warms quickly to the early spring, and as the thaw exposed the damp, protected earth, the primroses bloomed, their pale yellow joining the deeper yellow of the gorse flowers that had been hidden by the snow. The cats entered a new year.

Tom had been away most of the time, living a largely solitary life, finding food and shelter where he could. Peter had occasionally seen him in the wood between the farm and the village. Tom did not want for feline company, but he had no need of regular companionship. Perhaps the females did. Smudge and Pickle and Domino continued to share their ranges and the shelter of the barn.

As winter ended, Lucky was eight months old. He was still immature sexually, and had not yet grown to his full size. He was young, strong, active and increasingly self-confident. He had survived the winter and there was every reason to suppose he would go on to enjoy a long and full life.

The cats in the barn saw quite a lot of Shadow, whose persistence had brought him to a form of membership of the group. Once inside the barn he had refused to leave, and by now it appeared to be as much his home as it was the home of the farm cats. His presence was tolerated, and at times it seemed that the farm cats had even come to like him, although he remained nervous and would always run for cover if he was threatened.

The relationship between Shadow and the other cats had been intriguing. Although a coherent group of animals, such as the farm cats, may be bound by familial bonds, it must be possible for outsiders to enter, especially for the purpose of breeding. If this were not so, groups would be isolated and inbreeding would soon expose weaknesses among them. Males must be permitted to visit females – as Tom did. We might expect, though, that other non-relatives would be excluded, that even if an individual sought to join a group, its advances would be rejected. Shadow's advances were rejected at first, but not decisively. One possibility is that as an immature male, who was content to accept a subordinate role even to the kitten, Shadow presented so little threat that the cats felt no need to expend the effort that would have been necessary to eject him and prevent his return.

Many cat owners have direct experience of mutual tolerance between house cats that are quite unrelated, and it would be tempting to classify Shadow's relationship with the farm cats in the light of such commonplace domestic behaviour. To do so would be unwise. Two or more house cats that are brought together arbitrarily – so far as they are concerned – within a single core area of a shared range are placed in a situation in which closely related animals commonly find themselves. This may 'deceive' the cats into treating one another as they would if they were related. In the case of Shadow, no such deception was possible.

As winter came to a close, Peter's involvement with the project also finished. He had been offered an academic post in South Africa that he could hardly refuse, and the BBC, which as sponsor of the project had been paying his salary, agreed to release him a few weeks early. The film in any case was almost complete.

It was a sad parting. He had become fond of the cats and of the other friends he had made in Devon. He said his farewells, promised to keep in touch, and packed up the contents of his hut.

His last task with Maurice was to catch the cats. The time had come for them to move too. Like Peter, they had been temporary residents on Maurice's farm, so that he could film their day-to-day lives. Now they were returning to Oxfordshire, not to the farm from which they came, but to another permanent location where David Macdonald would continue to observe them and record their behaviour.

With care, one by one, the film stars were trapped and loaded on to the hired van that also carried Peter's worldly goods – including his books, scientific equipment and the tattered poster of Snoopy. His motorbike was strapped to the roof.

As far as their television public was concerned, the tale of Tom, Smudge, Pickle, Domino and Lucky was told. Mizzy, the Tibbles' pet ginger cat, was the only one left behind. After the others had gone, he visited the barn a few times, but to this day he behaves as if that wild bunch might come back.

The outside world

ONE DARK, cold, wet, windy night, David Macdonald sat in a hired van beside a main road near Newbury. He was waiting for the arrival of the van that was bringing Peter from Devon, together with his belongings and the cats that they both knew so well.

The second van arrived and the cases that contained the cats were transferred. Peter and David parted, David to return to Oxford and Peter to begin his long journey to South Africa.

It was Peter's work with the farm cats that had earned him his new appointment. At Oxford, David was friendly with a visiting professor from the Mammal Research Institute of the University of Pretoria. They discussed cat behaviour in general and the Devon cats in particular, and Peter was invited to South Africa to study a colony of domestic cats whose circumstances were very different from those of the Devon cats. The study would provide the material from which he could work towards a master's degree at the University of Pretoria.

The colony he was to study is on Dassan Island, a place with a hot climate, luxuriant vegetation, many cats, and one other human inhabitant, the lighthouse-keeper. It is a lonely place and less comfortable than it sounds. The new scientific recruit had to be able to endure solitude and discomfort, while conducting work with full scientific rigour. There was little doubt that Peter possessed the necessary academic qualifications and personal qualities. Even so, his task was not easy. The vegetation provided cover in which the furtive cats hid, and it was some time before he so much as saw one. He had to persevere. By the autumn of 1980 his project was completed and he was back in Pretoria writing his thesis.

His scientific work earned the admiration of his colleagues, but he won their profound personal respect too, and in a most unusual way. He spent his free time on the island learning Afrikaans, from books and from the lighthouse-keeper. The University of Pretoria is a pre-dominantly Afrikaans institution, and this is the language in most common use. On one of his visits to the mainland he walked into a seminar on animal behaviour at the University and joined in the discussion, in Afrikaans. It was the last thing that was expected from an English student – and a visiting one at that – and his popularity was guaranteed.

He kept in touch with David, as a friend but also as a colleague with whom he exchanged information and ideas about cats and their society, and he kept in touch with Maurice. The last Maurice heard of him, late in 1980, was that he had a new project, on the mainland this time, studying caracals. The caracal, or Persian lynx (*Lynx caracal*) is found over much of Africa. Rather larger than a fox, it lives on the grasslands and hunts animals up to the size of small deer and gazelles. The prospect of moving to a new species was exciting.

All this time the five cats from Devon were thriving back in Oxfordshire. David installed them in a habitat that was not too different from the one they had been used to in Devon. A friend owned land that extended into adjacent farmland, providing conditions rather like those on Maurice's farm; the garden even had an outhouse which could provide a substitute for the barn. David realised that the move from Devon could have disrupted the cats' social behaviour. To this extent, the transfer was something of a gamble, but it succeeded and the cats soon settled in their new surroundings. There was no discernible difference in their relationship before and after the move.

The purpose of the continuing observations was to monitor the long-term consequences of the social relationships that had emerged among the cats. In their new location David supplied adequate food for the entire colony, since the emphasis of the study had now changed.

After about two months, Tom began wandering again, as he had done in Devon, and his absences grew longer and longer. The cats were not fitted with radio collars, so it was not possible to track them as Peter had done in Devon. No one knows precisely where Tom went. He was seen now and then, up to about a mile from the colony, and he was known to visit some cats that lived on a farm about half a mile away. Probably he would come to no harm, but David tried to trace him, placing an advertisement asking for information concerning his whereabouts in the local post office, together with a photograph of the cat. It brought no result, however, and Tom continued to come and go as he pleased.

A new cat appeared on the scene. David named him 'Zebedee'. He was a mature male, ginger, large and ugly, and two or three years old, so at the peak of his condition. No one saw Tom and Zebedee meet, but

Lucky learning to climb trees, with Pickle

it was not long before Zebedee was seen more regularly than Tom.

Within Pickle's range there was a house owned by a lady from whom she was able to obtain food, so life for Pickle became very easy indeed.

Smudge and Domino became pregnant again, this time almost certainly by Zebedee. In the spring of 1979 Smudge produced four kittens and two weeks later, on 5 May, Domino also produced four. Once again Domino proved to be a poor mother and two of her kittens died soon after they were born. The other two survived and joined Smudge's four in a kind of merged litter that lived in the nest that both females shared. The females also shared in their upbringing, and although Smudge was the superior mother, Domino did improve.

The collaboration was a direct repeat of the behaviour that had been seen in Devon, confirming that the first occasion had not been an isolated occurrence. In neither setting had there been a shortage of suitable nesting sites or other environmental pressure that could have compelled Smudge and Domino to share a nest.

The collaboration shown by the farm cats is not seen in all cats. It appears that a whole range of social organisation may be found in cat colonies, varying between two extremes. At one end of the spectrum there is the full collaboration seen in the farm cats. At the other end there are cats that make individual, isolated nests in which their kittens are born and raised with no help from other cats. Between the two, there are colonies in which females make individual nests in which their kittens are born without outside help, but some time during the development of the kittens they are all run together in a kind of 'creche', where they are supervised by some of the females.

Other social differences have been observed. Among some colonies, for example, interactions among individuals are frequent and amicable. The cats seem to be good friends. Our farm cats related to each other in this way. In other colonies, which superficially appear similar, interactions are fewer and the cats appear much more as isolated individuals.

Why do some cats behave in one way and some in another, even when the environment is apparently similar in every important respect? David Macdonald and his students are working on this question, but it is possible that the basis of the difference is cultural. House cats occupy a core area that often contains no other cats. Cats that live independently of humans are in a very different situation. If they can collaborate there may be obvious benefits to the group, but there will be benefits to each individual as well, at least in the sense of 'you scratch my back and I'll scratch yours' – good turns may be reciprocated. Over many generations, therefore, it may well be that although collaboration makes no difference to house cats, among independent cats it is the collaborators who survive most commonly. If this is so, the difference in social behaviour between one colony of cats and another may reflect the time – measured in cat generations – for which the cats have lived apart from humans.

This is speculation, of course, and at best it is only one element that influences the social organisation of cats. A more important factor may be the way food is distributed within their ranges. Small sources of food, spaced at intervals, may encourage social interaction less than large sources of food obtained at one place.

By the middle of 1979 Lucky was an 'adolescent', and he helped to bring up Domino's and Smudge's six kittens. His help consisted mainly of being present to guard them. He never picked up a kitten or carried one, but he did play with them – far more than Tom had ever played with him. Lucky became a useful 'baby-sitter'.

Lucky – in the Oxfordshire countryside – now an independent young tom

Had he been a house cat, Lucky would probably have been sexually mature by this time. It seems unlikely that the rate at which he matured was strongly affected by his diet, for he was well fed. It is possible that the presence of Tom had an inhibiting effect, and that young males mature more slowly in the presence of mature males. House cats are usually isolated from older males and so they mature earlier.

As the winter of 1979–80 began, Lucky started to spray. Tom's absences may have removed the inhibition to his development, and by the spring of 1980 Lucky was fast becoming a typical tomcat. He was two years old before he was fully mature, and by then he was spraying often. He fought local rivals, and he was less tolerant of his relatives. He took to wandering off on expeditions, as Tom had done before him, and was away for days at a time. Local people saw him now and then while he was out on his forays. He had become as competent a hunter as his mother, Smudge. Lucky was a combative animal who could and did take care of himself.

Early in the summer of 1980, Smudge gave birth to one kitten. She had left the core of the colony and had made herself a nest about a hundred yards away, where she produced her kitten alone. This, too, was a direct repetition of the way Lucky had been born, but the career of the new kitten was sadly different from Lucky's. When it was about a week old, the kitten disappeared. David and his students searched and made extensive enquiries, but no news was heard of it and its fate remains a mystery.

At this stage, what had the project achieved? So far as David Macdonald was concerned, it confirmed his belief that cats are ideal animals for investigations of social behaviour, and that the ecological circumstances of farm cats provide an unusual and rewarding opportunity to study the fascinating aspects of cat society. Most of what we had known about cat behaviour previously had been derived from laboratory studies of animals living in very contrived circumstances. There had been several other important studies of domestic cats in the wild, but these had tended to concentrate on the relationship between the cat colony and the natural resources on which it depended – the ecology of the feral cat. Isolated colonies, such as those on islands or the one Jane Dards observed in Portsmouth Dockyard, are excellent subjects for studies of this kind, but the possibilities for studying social behaviour within them are very limited. The cats are too secretive, their private lives too secure from human scrutiny. Even in the dockyard, many important events went unobserved simply because they took place under cover. Farm cats provide a satisfactory compromise – the cats are unfettered, but relatively easy to observe.

Scientifically, much of the work in Devon had contributed a base-line for interpreting the 'language' of cats. It may seem trivial to devote so much time and effort to noting in detail the frequency and manner in which one cat rubbed its cheek against another, but such interactions as these form the basis of social structure among cats, and by close observation of them, in time it may be possible to construct a kind of 'cat grammar' for subsequent studies.

By the time the cats left Devon, a fairly clear picture had emerged of the way they had used the space available to them. Tom's range had been found to embrace several different farms and it was about ten times larger than the range of any of the females. The females had ranges that overlapped substantially, but within each of these ranges it is likely that there was an area that was not overlapped and which each female used more or less exclusively.

It was the collaboration in the birth and rearing of the kittens that provided the greatest insight into the structure of the cats' society. It was these events that created the greatest excitement, not only for David and Peter but also for the millions of viewers who enjoyed the television account of the cats' year on Maurice's farm.

The success of the domestic cat, and in no small measure its interest to scientists, is based on its great behavioural flexibility. It is not simply that its behaviour is not rigidly stereotyped, but that its behaviour varies during the life of an individual cat, so that the individual can accommodate itself to quite different social conditions. Lucky, for example, was brought up in a society in which he was well cared for, well taught, and encouraged to develop as a social animal, which he did. As he developed further, however, his contribution to the society, as a kitten-minder, came to an end. He became a leading figure in the society and then a disruptive one, until, at last, he began to wander off alone. He was entering a phase of his life during which he would be quite solitary, yet still a member of a society of sorts as he set out to establish a place for himself among the unattached males. The solitary phase, too, would end, for we may assume that he would find females of his own, interact with them socially as well as sexually, and so become a social animal again.

The cat is not unique in its adaptability, but its ability to live as a solitary animal, as a member of a feline society, or as a parasite on the fringe of human society is more highly developed than in most animals. In the future its continued success will depend on its ability to go on entertaining and commanding the respect of humans. There is no reason to suppose that it will fail, or that we or our descendants will tire of this kind of partnership.

If we are to understand cat society, perhaps we should alter our own viewpoint. Instead of seeing the cat as a part of the human environment, we should try to stand back a little and see the world as it appears to the cat, and ourselves as part of its feline environment. If we can do this, there is a contribution every cat owner can make to the more formal study of cat behaviour. The cat is the most observed animal in the world, but it does not require us to look for the unusual, or for behaviour we have not seen before. It requires us only to watch what is familiar, very ordinary, but to watch it perceptively, trying to discover how each small item of behaviour may have developed and how it helps the cat to succeed in our shared world.

There is no way we can harm the cat, or its reputation, by seeking to understand it. Indeed, the realisation that the domestic cat has developed forms of behaviour that have enabled it both to prosper in our homes and to hold its own among the wild fauna of our countryside can only increase our respect and admiration for it. Appreciation of the cat's own private world will increase our enjoyment of the times when it chooses to share ours.

Postscript

TOM, SMUDGE, Pickle and Domino were taken from Oxfordshire to Devon early in 1978, and in the winter of 1978–79 they were taken back again to Oxfordshire, together with Lucky. We have recounted their adventures as they settled into their new environment, and now it is time to end our story by bringing it up to date.

Tom was not found as a result of the advertisement in the post office. He was seen several times, but eventually he disappeared and never returned. He is presumed to be living in the Oxfordshire countryside.

Smudge produced one kitten in 1980, just as she had done in Devon, but soon after it disappeared so did she. A cat resembling her was killed in a road accident at the junction between a road and a farm track. The circumstances suggested that Smudge was the dead cat.

Pickle befriended the lady who fed her and before long moved into the house. Today she has demonstrated the adaptability of cats in spectacular fashion by abandoning entirely her feral existence. She is now a sleek, contented, house cat.

Domino has remained independent and still lives in and around the Oxfordshire hut to which she was taken from Devon.

Lucky had a serious fight with the male intruder, Zebedee. After that, his absences grew longer, and finally he disappeared and has not been seen since. Enquiries about him were made locally, and for a time people caught glimpses of him, but he never returned. He is presumed to have entered the solitary phase of his life and to be living in the Oxfordshire countryside.

Index

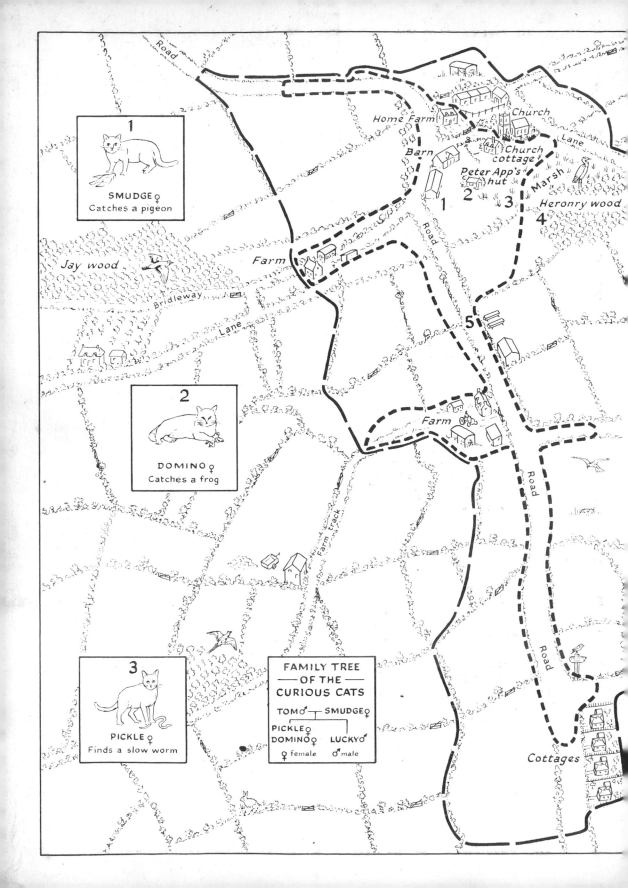

1 SMUDGE ♀
Catches a pigeon

2 DOMINO ♀
Catches a frog

3 PICKLE ♀
Finds a slow worm

FAMILY TREE
— OF THE —
CURIOUS CATS

TOM ♂ ⎯⎦⎡⎯ SMUDGE ♀
PICKLE ♀
DOMINO ♀   LUCKY ♂
♀ female   ♂ male

Road

Home Farm

Barn

Church

Church cottage

Peter App's hut

Marsh

Heronry wood

Jay wood

Bridleway

Farm

Lane

Lane

Road

Farm

Farm track

Road

Road

Cottages